NORFOLK & WESTERN

Electrics

By
Mason Y. Cooper

First Edition: AUGUST, 2000

Published by:

NORFOLK & WESTERN HISTORICAL SOCIETY, INC.
P. O. Box 201
Forest, VA 24551-0201

Library of Congress Catalog Card No. 99 076261
ISBN 0-9633254-9-3

Layout and Design By:
Greever Grafix
5080 Burns Road
Lilburn (Atlanta) Georgia 30047

Printed by:

Walsworth Publishing Company
Marceline, MO 64658

Introduction & Acknowledgements

Electric operations are an untold part of the Norfolk & Western Railway story. Existing in the shadow of the road's legendary steam program, it is understandable that its operations were overlooked. But without the benefit of electric traction in the early 1900's, the company would have faced far more difficulties in moving coal. Initial expansion through the barrier of Flat Top Mountain was done with a temporary feeder branch, which almost by chance became a part of the railroad's main line. Connecting to this branch on the western end of this feeder line the easily graded Ohio Extension was attached. This Ohio Extension eventually reached markets via Columbus and Cincinnati. Linking lines east and west, the original steeply graded and tightly curved Elkhorn coal line became a bottleneck to train movement. With mining in a high rate of production, little land or opportunity existed for right of way improvements. Instead, N&W sought out the best available motive power to move the heavy traffic... a trademark for which the road would become well known. Compounding these transportation difficulties was the ever-expanding number of coal loading facilities. As soon as one remedy was achieved, demand for increased service would overrun the solution.

By 1910, the level of coal traffic had outstripped contemporary developments in steam motive power. Concurrently, improvements in electric traction brought this alternate form of motive power to the attention of Norfolk & Western. Although several key features of the system finally selected were at that time untested, installation proceeded. Norfolk & Western's electrification of the Elkhorn Extension was the first A.C.-powered mountain railroad. The operation was widely studied and copied. This led to development of similar electric applications worldwide. Among the innovations, the system introduced the use of split-phase locomotives in the United States. These permitted higher operating speeds for trains on flatter portions of the road while reserving a lower speed-high torque setting for the Elkhorn grade. Additionally the electrification project was the first use of electric regeneration for train braking. Before this application, regeneration was largely a theory that had been attempted only once on a light European line.

Electric operations on the Norfolk & Western are an interesting tale. The system was pressed into the most difficult type of service on the railroad, overcame unforeseen teething problems, and prospered. Full planned development of the system was cut short by traffic shifts, a failed merger and finally economic downturn. The operation matured and quietly went about its duties, tucked away almost unnoticed deep in the Pocahontas coalfields. Eventually, mining activity abetted and allowed a right of way realignment which brought about an end to electric operations.

Norfolk & Western continued going about the business of moving coal while initiating an expansion of its railroad system. Events conspired to put the company back in electric traction in 1959 when it inherited a revitalized electric train operation through merger with Virginian Railway, capable of operating well into the future. The usefulness of this electric system was cut short by a change of operating scheme by the newly combined railroad.

This story is not light reading, but is intended to leave one with an understanding of the workings of N&W's particular operating system. This work was assembled using three sources of information. The Norfolk & Western Railway Collection, part of the Special Collections Section at VPI&SU, proved to be a most valuable asset. A good portion of the photographs used to illustrate this work are a part of this collection. The Interstate Commerce Commission Library provided material concerning railroad accidents. Finally, the Norfolk & Western Historical Society Archives provided material accumulated by its members as well as records recently deemed surplus by the Norfolk Southern Corporation.

Several individuals contributed time and effort to this project and have earned special recognition. Their individual efforts helped to fill in many gaps in this history and proved to be invaluable. Marty Swartz researched records located at Pittsburgh PA, home of Westinghouse Manufacturing, and provided construction records of the N&W Class LC-1, LC-2 locomotives, the powerhouse specifications, and material concerning the Virginian Railway. James Gillum researched the N&W Annual reports for electric material, and offered his photo collection for the project. Ed King provided antidotes from "Pokey Lore" concerning the electrics as well as period timetables which assisted in gaining a full understanding of operations. Ron Davis assisted in researching the N&W Historical Society Archives for pertinent material. Gary Rolih supplied maps and materials from his Traction Heritage Collection and turned up missing material from the N&W Historical Society Archives Magazine collection. Concurrent with the gathering of data, a series of articles on the Elkhorn Operation written by William Harvey appeared in the N&W Historical Society magazine *The Arrow*, and were helpful in pointing out additional directions for research. Janos Eros provided a background on Italian Electrification, specifically pioneer split phase AC operations. John McCoy provided material on the Vauclain and Balanced type steam compound locomotives. James Brewer provided the bulk of the proofreading, checking photo captions as well as text. He offered several suggestions that improved the story. In addition, Bob Bowers, James Cochran and Roger Nutting proofread portions of the text, and offered suggestions which were incorporated in the text. Jarrell Greever worked the many parts of this tale together into an artful and pleasing layout. Finally I need to thank the members of the Norfolk & Western Historical Society, without whose confidence and significant financial assistance, this book would not be possible. ■

This book is dedicated to the personnel of the Norfolk and Western and Virginian Railroads who served "under the wire."

FRONT COVER PHOTO: LC-2 2514 works another trainload of Pocahontas coal eastward from Iaeger, WV in 1926. The LC-2 motors, the most powerful of all N&W electrics, worked these duties for nearly 25 years.

BACK COVER PHOTO: A head-on end view of N&W Class C-2 No. 2514. The placement of the headlight and the dual purpose marker classification lights were, over time, applied to all electric locomotives.

NORFOLK & WESTERN Electrics

All photos are from the Norfolk & Western Collection, provided courtesy of the Virginia Polytechnic Instute and State University, unless otherwise noted.

ABOVE: A trainload of coal behind new Norfolk & Western Class C electric locomotives climbs the Elkhorn grade out of the Pocahontas field at Switchback, WV in 1915.

1 The Challenge

It was coal, or more specifically the movement of coal, that was the challenge. In May 1881, when Frederick Kimball saw Jordan Nelson using coal from a twelve foot outcropping to fire his blacksmith forge, he knew he had a tremendous source of traffic for his N&W Railroad. Here in the sparsely populated Appalachian mountains was a source of wealth. You did not have to be smart or dishonest to have a share. All you had to do was dig it up. For the most part, the seams (glossary) were so thick, they were over a man's head and easy to work. The coal cropped out on the sides of the ridges and as a result was free of flooding water.

An agreement was struck to bring N&W to the site, following a handshake deal between Kimball and the South-West Virginia Improvement Company. On March 10, 1883, the last spike was put in place. A rail line now snaked up Laurel Creek to the mine head. The area had already changed. Here was the new town of Pocahontas, constructed on a site known formerly as Powell's Bottom, 375 rail miles from Norfolk. Miners had 40,000 tons on the surface waiting to the moved. By the end of the year nearly 120,000 tons had been mined. For the railroad it became a game of catch-up. Overnight area farmers became mine operators as the rails followed the activity up the hollows. Options on lands were taken up by a newly formed Flat Top Coal Association. This firm which did no mining, leased land to mine operators, who in turn paid royalties on tonnage mined. The branch lines expanded.

At a point near Bluestone, VA (N&W milepost 374) where Laurel Creek emptied into the Bluestone River, a switch was installed for a line following that waterway towards the settlements of Bramwell and Duhring. Westward off the Bluestone branch, sidetracks snaked up Mill Creek, Simmons Creek and Flipping Creek towards the coal outcroppings on Flat Top Mountain. With the Bluestone River marking the eastern edge of the coalfield, three to four mining firms were each soon at work on these tributaries. Most of these, in the beginning at least, were small operations that grew in scope with the field. With the seam located on the hillsides, in watercourses

Construction crews work the first N&W train into Pocahontas, VA behind new Class H 4-6-0 106. Later that day the engine will pull the first loads of coal from the mines to Norfolk, VA beginning a series of coal trains which continue to this day.

known locally as "hollows," these were for the most part drift mine (glossary) operations. Workers walked into these mines to work the coalface, usually in parallel shafts to aid ventilation. Drift mines required no hoisting machinery. At first coal was loaded either directly into cars or a storage bunker, but soon was separated by screening into different sizes.

Near Duhring, 8 miles from Bluestone Junction, Robert Goodwill & Company, Louisville Coal and Coke, and Stephenson, Mullin & Company worked the two forks of Flipping Creek. Downstream, three miles from the Junction, a spur up Simmons Creek was shared by Caswell Creek Coal & Coke, William Booth and Company, and the Stephenson Mullin operations, whose rights spilled over the adjoining ridge line. On Mill Creek, John Cooper & Company worked rights to coal on the south side of the waters, while Mill Creek Coal & Coke Company retained rights to the north side. High up on Mill Creek and continuing onto the summit of Flat Top Mountain, Coaldale Coal and Coke Company worked a claim with a loader set in the depression formed by the run. Coaldale's mineral rights extended over the summit of Flat Top Mountain and onto the west slope. To better serve this claim, and open additional coal lands along the headwaters of Elkhorn Creek, additional rail construction was considered necessary.

Several routes came under immediate study for a mining spur utilizing a tunnel to reach into the Elkhorn region. The first route considered was via Simmons Creek, and a second by the Mill Creek branch. Each route offered specific advantages with Simmons presenting a longer westbound approach, and as a result an easier grade. Mill Creek offered a shorter route to the west slope, being located only three miles from Bluestone Junction.

N&W, which at first was pre-occupied by other matters, had tired of waiting for a connecting railroad to build into the coalfields from the west. Surveys began to open additional territory for mining, and perhaps locate a viable through route to the Ohio River. Work proceeded under the direction of William W. Coe, N&W's chief engineer. Speculation centered on a route west via Crane's Creek, located to the north of the immediate area. This effort was intended to thwart other railroads from entering the region from that direction. Kimball was wary of

Coke production was an important part of the picture in the Pocahontas coalfields. Substandard coal and dust went into coke ovens, such as those seen under construction here at Keystone, helping to maintain the uniform size and consistency of the product.

this alternative. Several individuals loosely associated with the company had attempted to form a syndicate to secure land options ahead of the railroad, which they intended to resell to N&W at inflated prices.

Meanwhile, back on the mining branch, a hard look at the Simmons Creek route revealed problems. Mining operations along Simmons Creek were in a more advanced state of development leaving less available room for railway right-of-way improvements. Also, a longer tunnel would be necessary at the summit to reach Little Fork branch of the Elkhorn as this waterway didn't advance as far east onto Flat Top Mountain as the South Fork of Elkhorn Creek. Finally, the route would be 2 to 3 miles longer than a Mill Creek line because of its location further upstream on the Bluestone River. Construction work was delayed for a number of other reasons, as the railroad was over-extended from the recent expense of getting its rail line from Central Depot (Radford, VA) into the Pocahontas coal region. And a shortage of rails had caused an increase in costs, which also delayed the project.

By 1886, when actual work began, the Mill Creek survey had been selected as the preferred route. In reality, no good way existed in crossing Flat Top. The Mill Creek survey offered the advantage of being located only 3 miles from Bluestone Junction to the mouth of the proposed tunnel. The grade was reduced by reworking the former mining spur. Surveyors staked out a line and removed the Mill Creek Coal & Coke loader to reach the site of the bore. On May 12, 1887, contractor Henry Fairfax, (N&W director and principle partner in the railroad construction firm of Fairfax & Diffendall) began digging a 3100 foot tunnel. It had been determined that whichever route was selected the tunnel would be dug in part through the Pocahontas coal seam which extended under the entire area. By working the tunnel through the seam, the digging would be easier. Coal is softer than the surrounding sandstone and shale, and could be sold to recover part of the expenses.

Concurrent with tunnel construction, N&W planned to rebuild the Bluestone branch to ease the eastern approach. Plans called for a new line to be built from a switch west of Bluestone Junction, with a short tunnel through Laurel Ridge. From here the line followed the west bank of the Bluestone to a point opposite the hamlet of Mill Creek where the route turned west up Mill Creek. The improved branch line lay to the north of the mining spur to the mouth of the east portal of the tunnel.

Work on the tunnel continued throughout 1887 by utilizing the pre-existing 3% mining spur to access the work site. A telegraph office was installed at the east end of the bore, to sort out construction traffic and maintain contact with the front office. Coal robbed from the work was shipped out. By the time of completion, 16,000 tons had been pulled out for Coaldale, which owned the mineral rights on the property. Boring pushed ahead with all possible speed, building on a descending grade of 1.4% from the mouth of the east end of the tunnel. By May 11, 1888, the bore, now known as Coaldale tunnel, was nearly finished with the completion of track laying.

At the west tunnel mouth, 4 miles from Bluestone

Mill Creek Coal & Coke's tipple occupied the hollow where Coaldale tunnel was built. The mouth of the future tunnel lay beyond the coal car on the center track.

Junction, a switch was installed in the line at a location known as Camp. This spot likely received its name during tunnel work. Camp was later changed to Coaldale. A spur curved southward from this switch to run along the west face of Flat Top Mountain to new coke ovens of Coaldale Coal & Coke. These ovens at first supplemented, then replaced the coal loader on the east end of the tunnel. Coaldale's operation held the most extensive mineral rights of the three original operators working on the old Mill Creek branch. The Flat Top Coal Association promoted the manufacture of coke (glossary), an excellent by-product for Pocahontas Coal. An effort was made to screen this coal, as it was somewhat friable (glossary) and subject to breakage on handling. The slack (glossary) removed went into coke production and the rest sent to market clean. Car supply was coupled to coke production with the general understanding the railroad would supply 150 new cars for each 100 coke ovens built. Coke production was not initially popular with some mine operators but it did much to insure uniform quality of product. As coal loading continued to expand, operations that were also involved in coke production received scarce cars on a priority basis.

Immediately on the west side of Coaldale Tunnel, the rails descended on a 2% grade dug out from the north side of the ridge carved by Cane tributary of Elkhorn Creek. The branch line was now known as the Elkhorn Extension. Little Fork tributary of Elkhorn Creek was crossed by a high trestle at the mining settlement of Maybeury. A little over six miles from Bluestone Junction, a switchback was installed to allow the rails to reverse direction westward and drop down to the valley floor. Here the mine loader and coke ovens of the Shamokin Coal and Coke Company were located on the south side of Elkhorn Creek. A spur split off at Maybeury running north up Little Fork tributary underneath the trestle from Coaldale to Norfolk Coal and Coke Company. The mine head was located east of the run, which fed both a coal loader and coke ovens at track side on the opposite bank. Back at Maybeury, the spur down from the switchback doubled back eastward nearly a mile to coke ovens and mines of Elkhorn Coal and Coke Company. A wye was installed, to allow engines to be reversed, with the tail running up the hollow of Webb Camp tributary where the coal loader was found.

The main stem of the Elkhorn Extension continued westward, dropping downstream. Seven miles from Bluestone Junction a wye was installed to reach the Lick Branch Colliery. This operation was set in the valley floor (or hollow) of that run and was the second in the Elkhorn basin owned by Norfolk Coal & Coke Company. It featured a horse-powered tramway across Elkhorn Creek from N&W rails extending from the minehead to the processing plant and coal loader. Ennis, a mining camp 8 miles from Bluestone Junction, was a busy spot. The loader and coke ovens of the Turkey Gap Coal and Coke Company were on the north side of Elkhorn Creek. Crozer Coal and Coke Company's ovens, which sat nearly in the stream, were fed by a tramway approximately a half mile in length to a minehead near Milepost 9 on the Elkhorn Extension.

Beyond this point the loader and ovens of the Houston Coal and Coke Company were located on the north side of the run at the mining camp of Elkhorn. A more developed mining settlement, Powhatan, was found at Milepost 11. The minehead of Powhatan Coal and

The first train through Coaldale tunnel pauses for its photograph before entering the bore. It is not likely that the passengers are aware of the smoky conditions awaiting them inside.

Class G 241 stands for its photograph under the spindly trestle at Maybeury in 1890. The bridge will get a steel replacement in a few years, the first of many as the as N&W increased capacity to deal with the volume of coal moving up the hill to market.

Coke Company was set in the hollow of the Trace Fork branch on the north side of Elkhorn Creek, feeding both a loader and coke ovens. Up to this point N&W's branchline had stayed on the north shore, but now crossed over to a tramway-fed loader of the Lynchburg Coal and Coke Company, then recrossed the creek to coke ovens belonging to that firm.

At the forks of Elkhorn Creek, located twelve miles from Bluestone Junction, the mining camp of Algoma was established. The forks of the watercourse marked the western most point on N&W's original Elkhorn Extension. A railroad wye was installed at the fork in order to turn the locomotives of mine runs. The tail of the wye formed a spur running up the North Fork of Elkhorn Creek to new properties of the Algoma Coal and Coke Company, and Gilliam Coal and Coke Company. Algoma's mine head sat in the head of the draw (glossary) for Buzzard's Creek while Gilliam's loader was further upstream along North Fork.

The newly opened Elkhorn Extension was inspected in November 1888 by President Kimball with a special train. A number of presidents and officials from other roads and assorted dignitaries were on board as guests of the N&W. Among them was railroad engineering consultant George Morrison, representing a well known Boston financier. He prepared a report, which was made on board in Kimball's private car. Morrison reported on the extent of the coalfield and progress in establishing mining. The report included a look at N&W's operations including the Coaldale Tunnel.

By 1890, twelve mining operations were in production on the west slope of Flat Top Mountain, while seven slightly older operations worked the east slope. The Elkhorn Extension, although laid out for light one way tonnage, was now the most productive branch line in N&W's coalfields. To accommodate the movement of coal, a number of improvements were undertaken outside the confined regions of the Elkhorn field. To consolidate mine runs into road trains, a coal assembly yard was established on the east of the mountain at Milepost 371. Named Flat Top Yard for the coalfield, cars were gathered into trainload numbers before continued movement eastward. Additionally, a reorganization of railroad operating departments led to the formation of the Pocahontas Division with headquarters newly established at Bluefield.

Looking beyond the Elkhorn, the possibility of a connection with a railroad from the Ohio continued as a goal of N&W. Having now reached the Elkhorn by way

In 1888, when N&W began planning a route westward, Bluefield WV was little more than a field and a few tracks. Within a few years, the area would explode with growth, becoming the headquarters for the Pocahontas Division.

of the Coaldale tunnel, the railroad had an obvious way to the Ohio by way of Tug Fork and the Big Sandy River. Still other routes first came under consideration. The shortest line and the first considered by N&W was a route northward from the Elkhorn through Walnut Bottom in Boone County, then down the Coal River to a connection with the Kanawha & Ohio Railroad at Charleston, or the C&O at St. Albans. The flaw with this survey would be a tariff division for coal traffic with other carriers. Secondly, the other carriers may have blocked westward expansion. These problems led to consideration of other options.

A second route, whose proposed location was published on a map in the Annual Report of 1886, departed from the end of the now extant Elkhorn Extension near Algoma. It ran in a northwestern direction up Burk's Creek over Indian Ridge to White Oak Creek and Pinnacle Creek to Pineville. This route continued northward through Elkins Gap and Walnut Gap in Boone County by Madison located on the Little Coal River, and on to the Ohio River. Crossing over the river to the east bank near Huntington, the road continued to Ironton, Ohio.

All work on a through route west was suspended until 1888, by which time the Coaldale tunnel was dug and the Elkhorn Extension was placed into operation. Also, the railroad's involvement in other internal projects had delayed progress on an extension to the Ohio River. During this time N&W was completing construction of a branch into the Cripple Creek iron region in Wythe County, Virginia, and finishing the Clinch Valley line which connected with the Louisville & Nashville. Additionally, ongoing adversarial negotiations with Richmond Terminal had resulted in failure to acquire the East Tennessee, Virginia & Georgia, an important connection with the N&W at Bristol, VA. These matters had occupied the railroad and by 1889 had been sorted out. N&W was ready to proceed.

At this time the previous surveys to the Ohio were abandoned. One reason for dropping these projects appears to have been the high operating costs associated with the crossing of four mountains on heavy grades. N&W remained interested in potential for coal loading in this region, yet recognized it's short comings as a through route. A completely new line was surveyed using the Elkhorn Extension as a jumping off point to reach westward. Following the Elkhorn Fork downstream to the Tug Fork of the Big Sandy River, which flowed into the Ohio, N&W had a route west.

Halfway down the river a problem presented itself. The lower reaches of the Big Sandy River lay outside of the coal region as well as did the Guyandot River, which paralleled it to the north. The Guyandot River valley had at one time been considered as an alternative route. Knowing this, outside interests had purchased options on certain tracts on the upper portions of this stream hoping to extract a high price from N&W. Kimball was aware of this, yet wanted the railroad to pass through undeveloped areas where natural resources waited to be developed. Kimball was depending on revenues generated from hauling line-side traffic to pay for construction. Since no coal

The Tug River afforded N&W easy grades for its Ohio Extension. Besides coal traffic, the region was rich in lumber looking for a route to market. The Kibler & Kay Lumber Company, typical of operatons at this date, was located in isolated country at Devon, WV.

existed downstream, the other business possible would be lumber. As these streams were navigable for flatboats, the country along streamside had been cut over of timber. This deprived the railroad of any potential source of traffic.

Between the two rivers lay the narrow valley of Twelve Pole Creek. Because it was not navigable and isolated, the timber along Twelve Pole laid undisturbed. It was reasoned that by crossing over from the Big Sandy to Twelve Pole, and following it west to the Ohio, both a railroad line and a source of revenue to fund construction could be less expensively located. The route of the Extension to the Ohio River was settled.

In May, 1889 stockholders voted at the annual meeting in Roanoke to proceed with this survey for the Ohio Extension. To affect construction, the company used the West Virginia and Ironton Railway Company, a separate corporation established for the purpose of building the line. This firm was leased and finally merged into N&W in 1890. Costs to build and equip the extension were estimated at slightly over 5 million dollars, while Kimball planned for potential additional overruns reaching another million and raised funds accordingly.

By March, the entire route was under contract. Work began in May of 1890 and by the following month there were fifteen contractors at work on the 191 mile route with over five thousand men. The unsettled nature of the country and heavy construction work proved to be most difficult. Only the occasional farmer was encountered, a portion of whom greatly valued their isolation. In some parts of the survey, no wagon roads existed, and contractors had to raft or pack in supplies by mule.

Costs of doing work proved so expensive that by September 1890, several contractors had abandoned their projects and slipped off at night to avoid creditors. At a bend in the river known as Hatfield Bend, the construction gangs ran head long into the Hatfield-McCoy feud. There was much shooting and lawlessness that the railroad had to cope with in addition to pushing construction. A national business depression existed, but the Ohio Extension proceeded onward throughout winter weather, with the intention of gaining a return on investment at the earliest possible moment. The Ohio Extension was now running well over budget.

By January 1891 roadway grading was complete for 116 miles and track laying was beginning to advance from both Elkhorn and Kenova. Many difficulties lay ahead for the company. On the survey there were eight tunnels and sixty-two bridges, including thirteen on the Elkhorn Fork before it reached the Tug Fork at Welch. The huge bridge over the Ohio was 1737 feet in length, rising 100 feet over low water. This was made necessary by legislation passed on behalf of the steam packet boats in order to clear river traffic. A 2149 foot approach over a steel viaduct was needed on the east side. To begin work, it was found necessary to wait until winter when the river was at it's lowest levels for coffer dam work around some of the piers. It was also necessary to build a connection from the recently purchased Scioto Valley & New England RR to the Ohio River bridge with the embankment at Salliday Creek taking longer than expected to prepare.

The financial strain on the railroad was severe. Although N&W had allowed for cost overruns, expenses were soon pushing the $8 million mark. Kimball was faced with the possibility of having to suspend work altogether, pay off contractors and carry interest charges for the work off of existing revenues, or continue construction. The decision was made to close the gap and complete the extension. In July, 1891 an additional $1 million was raised by stock sales to old investors. Still, by November of 1892, a floating debt of over 3 million dollars had been incurred.

Besides the debt, the railroad faced some challenging operating problems. Chief among these was the Elkhorn grade, onto which the Ohio Extension was attached. The

The first engine to cross the Ohio River, 4-6-0 No. 43, is seen at the construction camp at Kenova, WV. Soon, the line will be opened to give N&W a through route to markets for coal in the west.

stiffest grade on the line was in the coal fields of the original Elkhorn coal branch, 68 feet per mile. On the Ohio Extension, which was built with better grades as a through route, this amount of grade was only approached once for a short stretch near the summit of the Twelve Pole line. The original Elkhorn was ill adapted for its new role, but N&W would work for the short term with the difficulties. Curvature, although not excessive, was nearly constant in some parts of the route.

The final spike was put in place, joining the two halves of the Ohio Extension at 4 PM on September 22, 1892, midway between Hatfield and Pondmouth WV. The line hosted its first inspection train two days later, but much work remained to be done on the rough track before it could accept business. By November 1, 1892 the line was complete enough to allow it to be opened for traffic. Much ballasting, station work and other details still remained to be completed. But financially the Extension was breaking the railroad. Kimball wrote in the 1892 Annual Report, *"had the immediate results to the Company been alone considered, or the exceptional winter anticipated, the opening of the Ohio Extension would have been postponed until spring."*

It had been thought necessary to push construction to completion in order to bring in machinery needed to open saw mills and prepare for mining. Several months time would be necessary for installation of this equipment before revenue could be produced. N&W feared it would lose the whole season by waiting until spring. But the winter proved so severe that no work could be done, even with the line completed. In addition, the hard weather caused many hurriedly constructed embankments to slide or settle, bringing added expense to the railroad's bottom line. Operation of the Ohio Extension during the first winter was both difficult and costly. This added expense occurred just when N&W had planned to see the beginning of a return on its investment.

Back on the Elkhorn, other problems were adding to the woes for the railroad. Coal operators were not happy with coke and car supply. The coke ovens were idle during certain seasons of the year, limiting the potential mine production. Although N&W was continually adding cars, the expanding Pocahontas field was creating larger demands for coal cars. Inadequate track capacity in the Elkhorn mining region called for more passing sidings and spurs where space was already at a premium. Timber trestle bridges on the line were past due for iron replacements. Even the Coaldale tunnel contractor, Henry Fairfax, filed a claim against the company requesting additional payments. These were in excess of $23,100 to cover reported losses on the tunnel. Fairfax claimed the Pocahontas No. 3 seam of coal, which was robbed during construction to recover costs, had completely disappeared in places. As his contract stipulated additional payment if the coal seam got down to less than 4 feet in thickness, he now billed N&W for the short fall. After negotiation, the railroad denied the claim. Fairfax, a long time friend of Kimball who had first done construction work on the Shenandoah Valley Railroad largely on promise of payment, resigned from the N&W Board. Kimball had alienated a confidant.

While attempting to hold down costs, some immediate expenditures were necessary to handle business. The railroad constructed nearly 4 miles of double track between Norfolk and Lamberts Point, largely for the benefit of expanding coal business. At Lamberts Point, the servicing facilities were supplemented with an 18 stall roundhouse for locomotives required for coal trains.

Traffic from the Pocahontas coal fields had grown to such an extent that it was found necessary to double track the 10.97 miles from Bluefield to Bluestone Junction. At Bluestone, the earlier plans for a new line up to Coaldale tunnel from further up Laurel Creek were discarded. In its place a route with improved curvature was relocated to begin from a switch installed at MP 374 at Bluestone Junction. This line continued west on a more direct route by way of a 697-foot tunnel through the ridge. This plan utilized a bridge over both the Bluestone River and the newly established village of Cooper. The new route eased the grade to Coaldale tunnel with the route via the Bluestone branch remaining in service as an eastbound main track. Normal operating practice found westbound empties on the new revised line, while eastbound loads used the old Bluestone line. This gave N&W a defacto double track (although separated at some points) all the way to the east end of Coaldale tunnel at Ruth. Branch line mileposts on the old Elkhorn Extension were revised to reflect it's new status as a through route on the Ohio Extension.

By 1893 both expansion of mining and consolidation of operations in the Pocahontas Coal Field was taking place. East of Flat Top Mountain, spurs at Flipping and Simmons Creek pushed further up the hollows while small operators had either been bought out or thrown in with larger firms. West of the summit, in the Maybeury area at MP 379, Norfolk Coal and Coke expanded it's holdings along Little Fork. Mining operations continued much as before, moving westward along the N&W tracks until the former camp of Algoma was reached at MP 386. Renamed North Fork, the settlement no longer marked the end of activity.

Up on the North Fork (of Elkhorn Creek), mining advanced eastward uphill beyond the Gilliam Company loader with two operations setting up for business in Bear Wallow Creek, the new Rolfe Coal and Coke, and Rivermont Coal and Coke. Roanoke Coal and Coke, with its loader hard alongside North Fork, and Arlington Coal and Coke split the Flat Top holdings on either side of Toms Creek. Ashland Coal and Coke and McDowell Coal and Coke set up in the upper reaches of Winding Gap branch, while the holdings of Greenbrier Coal and Coke extended nearly back over the ridge into Maybeury.

Westward on the Ohio Extension from North Fork, WV, the Pocahontas field was opened to several new operators. Keystone Coal and Coke Company established operations at MP 387. The firm worked claims stretching along Burke Creek to a loader constructed at the mining settlement appropriately named "Keystone". More extensive were the holdings on the south side of Elkhorn Creek, which were assigned to Pulaski Coal and Coke Company.

Greenbrier Coal and Coke was located high up on the reaches of North Fork Creek at Greenbrier, WV. This view, although taken in later years, is fairly representative of the new style of mining operations established in the opening of the Pocahontas coal fields around the turn of the century.

This claim extended in a great arc from MP 385 above North Fork, to the wye below MP 388 and in a southeast direction up Dan's Branch. Pulaski's coke ovens and loader were set at the east leg of the aforementioned wye at the new settlement of Eckman.

This wye, with its tail extending southward up Dan's Branch Run, presented N&W with a suitable location from which to establish coalfield operations. The old wye at North Fork (Algoma) had grown too congested for this purpose, with increased area mining activity. No room existed for an assembly yard, with all traffic taken over the mountain to Flat Top Yard. The area around North Fork developed so rapidly the railroad was unable to set aside necessary land. At Eckman the opportunity to set aside a suitable tract presented itself at the beginning of mining activity. A yard was established at Eckman, extending east to west alongside the mainline and Elkhorn Creek. Dan's Branch, draining into Elkhorn Creek at Eckman, was home to several coal operators. On the west slope, fronting Pulaski's claim on the west side, Eureka Coal and Coke was set high in the hollow, with Shawnee Coal and Coke whose holdings extended down Dan's Branch to Elkhorn Creek and continuing westward to below MP 389 on the N&W main track.

Continuing on westward, now on the north side of Elkhorn Creek, three operations split up the holdings. The most extensive was Empire Coal and Coke which fronted on the N&W from near Keystone to its loader opposite Langraff at MP 390. To the west of Empire the coalfield was divided by the flow of Bottom Creek between the firms Bottom Creek Coal and Coke on the east and Tidewater Coal and Coke to the west. South of Elkhorn Creek the larger holdings of Peerless Coal and Coke stretched from near Langraff, to its ovens at MP 391 at Vivian. The Peerless claims ran 3 miles further along the south side of the railroad and Elkhorn Creek. For the most part the claim was only recently established, contributing little in the way of traffic. Further westward the country was most desolate. Only two settlements had recently been set up in the area: Norwood (later renamed Kimball) at MP 392, and the settlement of Welch at the confluence of the Tug Fork.

The continued financial well-being of the railroad rested largely on the revenues generated from these 14 miles of mines and coke ovens on Elkhorn Creek. The heavy load N&W carried included the costs and charges of 116 miles of new track making up the Ohio Extension. Complicating matters was an oversupply of coal on the commercial market. The newly opened Pocahontas fields on N&W, New River fields on C&O, and the Beech Creek fields on NYC all competed for too few customers. Although Pocahontas Coal was of superior quality, it was a relatively untested product. N&W engaged in rate cutting to attract new customers, further eroding railroad revenues. In May of 1894, lower market coal prices led N&W to seek an accommodation with coal operators to share the reduction. This effort was tied to a move to hold miner's wages at the current level in order to keep a strike, then underway in the northern fields, from spreading into the area. The action, although well intended and neces-

When the railroad reached Eckman, WV, the railroad set aside enough room for a coal assembly yard. This structure served as station and, for a period, yard office for the facility.

sary due to N&W's financial difficulties, did nothing but alienate the operators against it. Although a rate compromise was finally reached, it left too little return. On February 6, 1895, N&W entered receivership.

Other factors also contributed to the Norfolk & Western bankruptcy. In 1894 the new Southern Railway began diverting ETV&G traffic away from N&W at Bristol, VA, cutting into revenue. Besides this matter and the high costs associated with the new track of the Ohio Extension, N&W had maintained rates to mine operators which both brought high profits and encouraged development. Market conditions would no longer cover the returns N&W was paying, leading the operators to reluctantly accept new terms from the railroad. In turn the operators attempted to cut mine wages, leading to a 4-month strike in 1895. This caused much hardship for all parties, but was eventually settled.

The improved coal rate structure, along with an increase in loadings from the Pocahontas fields as they continued to develop, pointed the way out of the N&W receivership. Every year coal and coke loading increased. By March 12, 1896 the receivers were able to issue a plan for an income based reorganization based on projected earnings from coal and coke. Implementation of the plan was delayed by the coming of a coal strike and by unsettled conditions in the security market. The railroad changed its name to the Norfolk & Western Railway and assumed operation of all property, except the Scioto Division and the Roanoke & Southern on September 30th. N&W added these latter two parts on December 1, 1986.

The new N&W "Railway" faced the same old problems and competitors, earning a living on the small difference in coal rates between mine and market. The national economy began recovery in 1898, at which time the demand for coal equaled supply for the first time in a number of years. Recovery in the coalfields was slower, in part because of the failure of the railroads to work together to avoid undercutting of rates, and because N&W's coal operations were expanding in size as new areas of the Pocahontas fields came into production. By 1899, traffic was beginning to recover, although the rates received for coal had risen only marginally. Various traffic pools and associations were tried to bring about a further increase. Gradually, the interests of the railroad and the producers drifted apart in their efforts. Mistrust existed, partially because rates were not posted publicly nor adhered to when market conditions changed.

By 1900, traffic began to stabilize, largely because of a "community of interests." This was brought about by the purchase of controlling stock, in the case of N&W, by the Pennsylvania Railroad. By the end of 1901, PRR had acquired 39% of the railroad's stock, enough to seat several directors. In what at the time seemed an unrelated matter, a strike in the northern anthracite fields in 1902 drove many customers to their first exposure to the superior qualities of Pocahontas coal. Finally, in 1903, the Elkins Act called for the public posting of rates and prohibited general rate cutting. As a result, rates became stable ending a period of uncertainty as operators were able to see that they were being treated equally.

As a result of these events a tremendous volume of traffic hit the N&W almost overnight. Although some of this coal traffic went west, the greatest markets at that time were located to the east. These were the manufacturing centers then in existence in New England and around New York City. The largest amount of coal moved to Norfolk where it was then either shipped or barged to market. Prior improvements already made by N&W (the pier at Lamberts Point, double track and coal assembly yards) were now paying big dividends for the road. The railroad was moving an average of 39 coal trains eastbound a day. The stiff Elkhorn grade and cramped Coaldale tunnel, both originally constructed for light branch line operations, were now bottlenecks to operation. By the following year the demand for Pocahontas coal was the greatest ever known. The railroad had gone from feast to famine, then to feast again. Dealing with the movement of coal would, from this day forward, be the driving force for innovation and improvement on the Norfolk & Western Railway. ■

N&W planned improvements to earn maximum benefit from growing capacity in the coal fields. To this end, first one, then a second coal pier at Lamberts Point in Norfolk to keep up with the increasing stream of coal. This photo shows Pier 3, added in 1900, in its original configuration. It was constructed as a gravity dump pier, then later converted to electric operation.

2 Searching for a Solution

The newly organized Norfolk & Western Railroad had been in existence two years when the Pocahontas coalfields were first discovered. The financial backers of the railroad were keenly interested in development of mineral deposits along the right-of-way. This influence led the railroad to construct its right-of-way accordingly, in order to handle heavier than contemporary car loads, and to keep abreast of the latest, most powerful motive power.

At that time the 4-6-0, or ten-wheeler, was the standard freight locomotive. This was also a period of great advancements in design when a top line engine could become hopelessly out of date in five years. It was quickly recognized that the newly developed 2-8-0, or consolidation type, was better suited for coal traffic. It had a larger proportion of its weight carried by the driving wheels, making it capable of hauling large loads, yet would track well when working over the roadway. This would be a better suited design for N&W's evolving coalfield operations.

The Elkhorn region soon eclipsed other coal branches both in terms of loads and difficulty of operation. Empty hopper delivery trains began the operating day at Bluefield. The trains worked westward heading first downhill to Graham, junction of the Clinch Valley line, then westward and up to Flat Top Mountain and through the tunnel. The line then dropped down again on the steep grade into the Elkhorn valley. Here the delivery train began setting off hoppers as needed at mines. Loads were picked up to be relayed with the remaining empties westward to the yard at Eckman. The remaining empties were set out for relay to other mines and the loads assembled into an eastbound movement. The loaded eastbound train then worked up the Elkhorn valley, through the tunnel and down to Flat Top Yard, where the trains filled out with additional loads off the Pocahontas and Bluestone branches and traffic forwarded out of the Elkhorn field

At the opening of the Pocahontas coalfields in 1883, the Class H 4-6-0 was among the best N&W motive power assigned to freight service.

by previous trains before moving east. At Bluefield, the loads were inspected, reclassified and weighed. Crews were changed and the trains were relayed to tidewater.

The first of the more powerful 2-8-0 steam locomotives was received in 1883 concurrent with the opening of the Pocahontas coalfield. Designated Class I, the engines were nearly 7 tons heavier than the Class H 4-6-0s. The new Class I 2-8-0s used slightly larger 20" x 24" cylinders and 50" driving wheels, a decrease in 5" over the Class H engines, resulting in a boost in tractive effort. Ten locomotives were received from Baldwin in 1883-84, to which 9 from the Roanoke Machine Works were added in 1884-85. These locomotives handled traffic off the Pocahontas and Bluestone branches, then the beginnings of the coal field.

In road service, the 2-8-0 consolidation on N&W proved to be an excellent freight locomotive. The railroad acquired a total of 68 of this type from Baldwin and Roanoke Machine Works, beginning in 1886. Using the same cylinder dimensions and 50" drivers, the new Class F engines weighed in at nearly one ton less than the older Class I power. Excepting three faster Class L freight 4-6-0s delivered late in 1885, delivery of the Class F 2-8-0s continued into 1887 giving N&W enough of these locomotives to re-equip freight service on other parts of the system.

By 1887 the railroad had decided to proceed with the Elkhorn Extension, and to this end had begun digging Coaldale tunnel to open the area. In planning for the steep Elkhorn grade and the increasing coal traffic, N&W looked for a more powerful 2-8-0 design to operate on these lines. In December of 1888, the Roanoke Machine Works completed the first of the new Class G consolidation. The locomotive incorporated the common 20" x 24" cylinders and 50" drivers which had already proven themselves on the railroad. This allowed an economy by utilizing common parts, which could be interchanged among the 2-8-0s. The new Class G 2-8-0s featured a Belpaire firebox. Few other railroads excepting the Pennsylvania RR used the Belpaire design, although the experts believed it superior to the standard rounded firebox because it provided better locations for bracing and support. The firebox was installed on a larger boiler, weighing in at nearly 7 tons more than previous consolidations. Beginning in 1889 and continuing into 1893 both Baldwin and Rogers delivered 125 more of these locomotives. This was enough to equip the recently completed Ohio Extension.

The N&W Railroad was aware of advances underway in steam engine design and followed the debate raging in the technical journals of the day. Baldwin was at this time a leader in locomotive construction. In that

Class I 2-8-0 125 pauses during its duties for a photograph in March 1885. The 2-8-0, or "consolidation," quickly gained favor in freight service as it had a greater proportion of it's weight carried by the drivers than the 4-6-0.

The Class G 2-8-0 shared many common parts with the Class I but with greater steaming capacity, it provided more power. This proved to be a valuable asset on the stiff Elkhorn grade.

era, the cylinder saddles were cast in two halves and bolted together. Baldwin did excellent work in preparing its fittings, which resulted in fewer steam leaks and reduced shop time. N&W favored the firm with its orders. Baldwin promoted a four-cylinder compound-type design, which was patented by it's official S.M. Vauclain, in the debate then in progress between Cross, Balanced and Mallet type compounds. N&W was keenly interested in these locomotive developments, and looked to them as a way by which it might reduce operating costs in its Elkhorn coal field. The Vauclain compound promised two advantages. First was the need for the low pressure cylinder to have a supply of steam readily available when needed. When in operation, if the high and low pressure cylinders are not synchronized, as was the case with the Mallet design, then the low pressure cylinder can starve for steam. (This was the main point the other designs stressed in selling arguments against the Mallet.) The Vauclain compound solved this problem by having the high and low pressure cylinders locked together. Although both the tandem and cross compounds also did this, the steam distribution on those systems suffered, according to Baldwin, by both distance and restriction. Even with the cylinders locked in phase, steam distribution from high pressure cylinders to low pressure cylinders suffered if the volume of the passage way between the cylinders was large as compared to the volume of the high pressure cylinder. In other words, the steam expanded and lost a portion of its energy in the passage chamber between the high and low pressure cylinders.

Beginning in late 1889, the first Vauclain compounds were placed into operation on the Baltimore & Ohio Railroad. Baldwin promoted the Vauclain's use on the B&O and succeeded in placing the design on several other railroads in both freight and passenger service. In 1893, N&W embraced the design and purchased 31 Vauclain compound freight 2-8-0s. These Vauclain locomotives were all delivered to the railroad by 1894, featuring a 14" x 24" high pressure cylinder underneath a 24" x 24" low pressure cylinder and rode on 56" drivers. Assigned to Class T, the locomotives were noticeably larger, weighing in at 10 tons more than the Class G engines.

On the Norfolk & Western, the Vauclains were a failure. The problem was that the force exerted by the two cylinders was unbalanced at low speeds. The engines worked well at high speeds and early cutoff, but on the steep Elkhorn grade the low-pressure cylinder exerted considerably more force than the high pres-

Compound cylinder steam locomotives promised savings through a reduction in operating costs. N&W invested its resources in a large fleet of compounds for both freight and passenger duties. Class T 2-8-0 321 is a Vauclain compound, which failed to live up to expectations.

The Class G1 locomotive carried a working boiler pressure of 185 lbs, and No. 7 was originally numbered 353 and ordered for pusher duty on the Elkhorn grade. The engine was sold, then later re-absorbed by N&W and is today displayed at Bluefield, WV.

sure cylinder at slow speeds and long cutoff. This caused the cross head to twist, which then worked apart causing the engine to fail. In addition, the demanding loads and operating conditions accelerated wear. Any savings gained from reduced fuel and water consumption were eaten up by heavy maintenance charges. The constant high steam demands of the heavy Elkhorn grade required a large steaming capacity. The Vauclain firebox was simply not large enough to supply the cylinders. Bushings were applied with a smaller piston to reduce cylinder size and steam requirements to no avail. But most importantly, problems with the steam distribution system could disable a locomotive bringing the entire railroad to a halt. These repair expenses along with the expenditure for motive power that had promised a more economical operation came on the eve of the railroad's receivership, and N&W could ill afford them. At a later date the troublesome Class T 2-8-0s were rebuilt with new steam chests and simple cylinders.

A small order for a proven design, six Class G engines, was all the N&W could afford in 1896. These were received from Baldwin Locomotive Works, and supplemented the pool of engines following the disappointing performance of the Class T 2-8-0s. The following year, advances in metallurgy led to the addition of seven "improved" Class G engines to the N&W roster. Now boiler pressure could be increased from 150

The Class W 2-8-0 featured larger drivers and were more powerful than the Class G engines they replaced. On the stiff Elkhorn grade, the limited steaming capacity of the engines quickly became a handicap. Despite this, the engines served long lives in other duties. Here, W1 813 and its crew pause from their yard assignment for a photograph.

lbs. to 180 lbs. which rendered greater pulling capacity. These seven 1897 Baldwin engines, designated Class G1, were assigned as Elkhorn helper engines where the additional steaming pressure could be used to maximum benefit.

Still, N&W continued to show interest in savings promised with compound cylinder locomotive designs. The potential for the reduction of operating costs was an argument the railroad could not withstand. Despite the failure of the Vauclain, Baldwin continued to refine the design as well as market other compound cylinder engines. Baldwin's efforts led to a balanced Vauclain with the two smaller high pressure cylinders placed inside the frame, with the larger high pressure cylinders outside, as with conventional locomotives. Baldwin promised reduced maintenance from this advance. Steam distribution was controlled by a single piston valve, but N&W, recalling it's prior experiences, rejected the engine, instead showing more enthusiasm for a competing two-cylinder cross compound design.

Between 1898 and 1904 the railroad purchased ten of these new cross compound 2-8-0s from Baldwin, designated Class B. In these two cylinder engines, steam was admitted from the throttle directly into the high pressure steam chest passing through the saddle of the high pressure cylinder. The low-pressure cylinder was nearly the same as an ordinary simple engine, except that its steam came from an intermediate receiver called an "intercepting valve" from the high pressure cylinder, rather than the dry pipe. Besides the cross compound cylinders, other innovations such as extended piston rods were applied. These rods protruded from the front of the cylinders and were intended to help support the pistons. In theory, this reduced wear on the piston rings, reducing the frequency of shopping. The Class B locomotives were the largest of its kind on the railroad, weighing nearly 33 tons more than the reliable Class G engines which had so far carried the bulk of traffic for the railroad.

In time the Class B compounds proved to have their own unique troubles. Quite a bit of skill went into making the intercepting valve function to balance the effort between the high and low pressure cylinders when working simple. It is an understatement to say that the Class B compounds were difficult to get started. This was especially worse in the Elkhorn field where full power was needed from the beginning and on nearly all parts of the climb up the grade. Depending on the receiver and volume of steam, there would be certain speeds where the low pressure cylinder would starve the high pressure cylinder. Only an engineer with a good deal of experience and full knowledge of his particular engine would know what reverse setting would work well at any certain speed. Enginemen could get at

The crew of W2 No. 683 stands by for its portrait. With a larger boiler placed above the drivers, the W2 Class corrected many of the shortcomings of the W 2-8-0.

steam distribution valves and block them accordingly to operate in simple. This was done without supervisory approval simply to get over the road more easily. This negated any savings from compound operation. With the locomotive operating in "simple" the firebox was unable to produce enough steam for the tough Elkhorn grade. When the advancement of superheating proved itself, the railroad considered the small advance of power from cross-compounds not to be worth the added expense. They were rebuilt with simple cylinders. The Class B locomotives remained the largest 2-8-0s on the railroad.

Twice the railroad had tested compound locomotives, both times coming away with unsatisfactory results. The recovering demand for coal was catching up with the abilities of the proven Class G design. Not enamored with the cross-compounds, once again N&W returned to a more conventional simple-cylindered 2-8-0 design with the arrival of thirty Class W engines from Baldwin beginning in 1898-99. These were followed in 1900-01 by 45 similar locomotives, initially designated Class W1, on account of the substitution of the slide valves with piston valves.

Up to this point, a saturated steam locomotive with slide valves had proved superior in performance to a saturated engine equipped with piston valves. As superheating proved its worth, problems began to occur with the slide valve motive power which changed this dictum. The superheater originated in Switzerland and raised the temperature of the steam by routing it back through large tubes in the boiler before relaying it to the cylinders. The increased temperature of the superheated steam led to excessive scoring of the slide cylinder valve plates. Piston valves eliminated this problem and began to appear on new locomotive purchases. Still, the retrofitting of locomotives proceeded piecemeal due to the added expense of the cylinders. In general, superheaters were applied to older engines on a case by case basis when it became necessary to replace the cylinders.

The 75 newly arriving Class W and W1 engines were much larger than the Class G 2-8-0s, taking advantage of advancements in the rail industry. The bigger proportions allowed the use of 56" drivers and 21" x 30" cylinders. Boiler pressure was increased to 200 lbs., with a deep narrow firebox setting between the rear two pairs of drivers. In operation in the coal fields it soon became apparent the firebox on the Class W 2-8-0s did not provide sufficient grate area for steam production under constant full power applications required in the Elkhorn region.

In October, 1901 the first of an improved design, classified W2, was released from Roanoke Shops. The locomotive featured a wide firebox which sat above the rear set of drivers, increasing its weight by 5 tons. This arrangement provided about 25% more area for coal combustion. Out on the Elkhorn grade, the W2 2-8-0 quickly proved its worth. Soon large numbers of W2 locomotives could be found on the hill. The railroad eventually acquired or built nearly 200 of these locomotives and they arrived just in time.

In 1902, the demand for Pocahontas coal began to skyrocket and N&W found itself scrambling to keep up. Since completion of the Ohio Extension in 1892, more mines had slowly been developed. The greater part of N&W's coal was destined for eastern markets and that meant movement up the steep valley and through the cramped Coaldale (or Elkhorn) tunnel. Eastbound coal train assembly took place at Eckman yard. By the turn of the century the average train was twelve hundred tons. These were handled by a combination of three Class G or I 2-8-0s, with one on the head end, a second cut in mid-train, and a third pushing. It was necessary to have the second engine cut in mid-train, rather than on the front or rear, to cut down on crew exposure to the noxious gases and heat inside the confined Elkhorn tunnel. The introduction of the Class W 2-8-0s gave the railroad additional options in dealing with this operating dilemma. Two of the larger Class W locomotives possessed more tractive effort than three of the smaller 2-8-0s. Depending on coal traffic volume the railroad could either eliminate the mid-train helper or increase tonnage handled on each train.

Engineer Alfred H. Hock, in train service out of Bluefield, later wrote of his operating days on N&W's Elkhorn grade in a two-part article appearing in *Railroad Magazine*. His description of what it was like to operate a train on the Elkhorn has left us a colorful picture of this difficult job.

Hock wrote... *"Ennis, at the foot of the mountain, was the beginning of a single track which extended to the top. And at the top was Elkhorn tunnel, nearly a mile long, having a fairly heavy grade through it. The tunnel had no artificial ventilation. It was a veritable 'Black Hole of Calcutta.' When the men started through on engines they would cover their heads with jackets or they would wet a bunch of waste and put it over the mouth and nose. Exhaust from the engines, with smoke and gas was terrible. Often the men came out gasping for breath. Their clothes felt as though they were on fire..."*

Hock recalled an unfortunate incident which occurred on Sunday, April 19, 1898. He had been called to work for pusher duty and by an error in crew assignments another man got this job. He was switched to a later-running Elkhorn turn. An Elkhorn Turn was a train called at Bluefield and consisted of empties for the coalfield. It would work west through the tunnel and down into the Elkhorn valley supplying empties to the mines, then returning with loads from the yard at Eckman. Hock reported for duty at Bluefield and worked his way west.

Hock continued.... *"When I stopped at Ennis, there was gloomy news. The train for which I'd first been called had stalled in the tunnel. Head engineer Kennedy had been carried out dead and several trainmen had been overcome with fumes. Engineer Honaker (on Hock's job) had stayed in the tunnel until he could stand it no longer. Then he reversed his engine, breaking off several cars from the train and backed out. Fortunately the train did not follow him back. If it had, there would have been a tragic runaway."*

By 1900, traffic had grown so heavy that the railroad was operating an average 39 coal trains a day up the single track grade and through the cramped tunnel. The total of all train movements, including empty hopper, helper, freight and passenger moves for a typical day could reach 92. During the summer it took from 17 to 55 minutes, depending on the wind, for the smoke to clear from Elkhorn tunnel. That is, if it cleared at all! Heat inside the bore kept the temperature 30 degrees hotter than the outside air.

Two other incidents took place in 1900 with serious consequences. An eastbound coal train stalled in Elkhorn tunnel, but managed to get itself restarted and was moving ahead slowly. Another coal train following behind disregarded the block signal indication at Coaldale, as it evidently attempted to make a run through the bore. Just outside the east end of the tunnel at Ruth, it collided with the first coal train running ahead.

A second incident that year occurred when another eastbound coal train stalled in Elkhorn tunnel, this time breaking in two. Three tries were made to get the coal train restarted. While trainmen were making a third attempt to re-couple the cars, a brake failure occurred. The rear twenty cars, along with G1 2-8-0 helper engines 257 and 563, ran away backwards. Out of control, the runaway picked up speed and at a sharp curve east of the bridge at Maybeury, locomotive 563 derailed. No one was killed, but it was found the engineers on the two locomotives had nearly asphyxiated in the tunnel.

Other mishaps occurred at Coaldale causing the

railroad to rethink the wisdom of building through the coal seam. In November, 1898 a derailment at the west portal of the tunnel almost caused a fire which nearly ignited the coal seam. Although the fire train was called from Bluefield it was the quick action of employees on hand who got the fire under control before it spread.

The arrival of the powerful Class W2 2-8-0, with its higher tractive effort, meant the railroad was able to temporarily eliminate the midtrain helper. The typical coal train now carried 1200 tons with a W2 2-8-0 pulling and a second W2 pushing. By 1902 the demand for Pocahontas coal had taken off. The coal had become the favored fuel of the U.S. Navy, as well as many commercial carriers. The railroad was again hard pressed to meet demand.

On busy days, there were as many as 57 eastbound coal trains up and over the Elkhorn. The conditions inside the tunnel worsened and had now become unbearable. High levels of exhaust fumes and water vapor made breathing nearly impossible. Suffocation was a threat not only for the many coal train crews, but a danger on the unsuspecting passenger trains.

Charles S. Churchill, chief engineer for N&W, looked for solutions to the problem by studying similar operations on other railroads. Looking to Europe, both the Italians and Swiss faced a similar problem with rail tunnels in the Alps. Success had been obtained for ventilation by forcing air through the bore to clear out gases from locomotives. The initial trial was a four-mile long tunnel at Pracchia, Italy, followed by an application on a longer bore at St. Gotthard, Switzerland. Churchill liked the idea, deciding to proceed with modifications to Elkhorn tunnel to install a forced ventilation system.

Further refinement of the European system was necessary in order to obtain the most effective use. No similar operation then existed in North America. N&W's tunnel, because of the tight clearances, needed physical modifications before such an installation could occur. It was decided to work at the Coaldale portal, on account of the prevailing winds. Here the cross-section of the bore was enlarged at the tunnel mouth by about eight feet. This enlargement tapered back working into the tunnel about 52 feet, to the original cross-section of the tunnel. This additional space formed the ducts through which air would be introduced inside the bore. A shell was made using structural steel angles, and wooden sheaths were constructed to the size of the original tunnel, running thirty five feet in length. This was installed at the west end of the bore, leaving a space of 17 feet between the end of the shell and where cross-section work inside the tunnel ended. This space was used as an air passage. At the west end of this enlarged section, on each side of the portal, two large fans were installed just outside the bore.

Charles S. Churchill served as engineer of maintenance of way for N&W, and with fellow employee Charles C. Wentworth, invented the forced air ventilation system used in Elkhorn tunnel.

The two fans forced a jet of high velocity air into the tapered spaces. The resulting draft created a low-pressure area just inside the mouth, drawing additional fresh air through the tunnel entrance from the west. By using this procedure, no door was necessary at Coaldale to assist with clearing. This allowed air to be forced into the tunnel even while an eastbound train was entering the tunnel. It was found advantageous to use the fans while eastbound trains were in the tunnel, with the force of the air moving the smoke from the locomotive along ahead of it faster than the speed of the train.

First public announcement of the project appeared in the August 14, 1900 edition of the *Bluefield Daily Telegraph*. The railroad laid out the plans, and the newspaper congratulated the railroad on taking steps to improve the conditions in the tunnel. A report on the installation of the fans appeared the following March, when it was reported a labor force of 50 to 60 men had been at work on the project for several months. By this time the job was nearly completed. The total cost was figured to be as much as $75,000.

The forced air ventilation system was based at the west portal of the tunnel at Coaldale, WV. This view, taken in 1917 after the electrification had been installed, shows the equipment when the fan motors were powered by coal-fired steam boilers. Later the motors were converted to electricity.

Additions were made at Coaldale to support the air-handling equipment. Prior to this time, a busy coal mining camp existed here, with a switch installed leading to coke ovens located to the south of the railroad on the western side of Flat Top mountain. N&W's presence formerly consisted of a roadway section foreman's house and a train block operator's station. A switch was added to these facilities, installed at a location on the north side of the tunnel. This became a stub-ended siding from which coal could be spotted for the powerhouse. The coal was used in the boiler to generate steam for two 75-hp engines. These in turn powered the tunnel fans, the wheels of which were 14 feet in diameter. A 300-foot well, drilled by a contractor, supplied boiler water. A prominent feature of the installation was the boiler smokestack, which was later raised in order to increase draft.

The fans were placed into operation on April 1, 1901. A report in the April 9, 1901 edition of the *Bluefield Daily Telegraph* reported bad news of poor performance. The fans were unable to force air all the way through the bore. The malady was caused by not running the equipment fast enough. Tests were conducted, leading to an effort to speed up the fans by 5, 10, then 15%. The April 22, 1901 edition of the *Daily Telegraph* reported the system now cleared the tunnel of smoke as planned. Later that year an article in *Railroad Gazette* appeared which reported, "*....the good report of performance of* [the Churchill-Wentworth ventila-

This postcard shows the stack extension on the tunnel powerhouse at Coaldale, WV. The extension was necessary in order to increase the draft on the boilers, enabling them to become more powerful. The top of the old stack appears as a ring around the structure near the top.

Roger Nutting Collection

tion system] *as published in our earlier issue has been confirmed by further experience. We are informed that people on the line, going into the coal field, now regard the trip through the Elkhorn tunnel the coolest part of the trip. Car windows are left open and no attention is paid to the fact that the train is passing through a tunnel."*

The arrangement and operation of the forced air system was improved over installations in Europe. This led Charles S. Churchill and his co-designer Charles C. Wentworth, a N&W bridge engineer, to seek patents. Eventually the Churchill-Wentworth system was installed on Pennsylvania Railroad's Gallitzin tunnel west of Altoona, PA, Chesapeake & Ohio's Big Bend tunnel near Talcott, WV, and the Virginian Railway's Merrimac tunnel.

N&W had seemingly arrived at the solution to its problem, excepting for one circumstance out of its control. Each year eastbound coal movements continued to rise. More loads in bigger cars and increasing numbers of trains appeared on the Elkhorn grade. Excepting a depression in 1907-08, the demand for Pocahontas coal rose each season. Coal traffic had again outgrown the limits of the motive power and N&W looked for something with more capacity than the W2 2-8-0. The railroad had continued to sample the cross-compounds, purchasing one final Class B locomotive in 1904, and had added Class W2 2-8-0s to the roster as late as 1905. The W2 had reached the limits of its steam capacity when handling the heavy coal trains on steep Elkhorn grade. That year the railroad began to look for a locomotive with larger grate area and more boiler capacity. The search would offer more than one solution.

The railroad studied what was used by other lines operating in mountainous regions, and other roads handling coal and iron ore traffic. In California, they

The Class M 4-8-0 was developed to keep pace with traffic in the growing Pocahontas coal fields. Traffic increased each year to such an extent that each new design was hopelessly out of date within a short period of time.

The 4-8-0 quickly became N&W's standard locomotive for coal train service. Class M 459 rolls a train along with brakemen riding the loads, ready to apply the hand brakes if needed.

discovered that Southern Pacific had used a 4-8-0, or twelve-wheeler, for a number of years in the Sierra Mountains. The locomotive had a number of features that N&W particularly liked, in that the 4 wheel lead truck was able to support a heavier boiler without increasing weight on drivers. This meant the engine could be introduced to service without having to strengthen or replace existing bridges. Too, the short wheelbase meant it would fit on current turntables. It had a higher tractive effort and had been used successfully in ore service on Great Northern and Duluth & Iron Range.

The 4-8-0 was adopted as N&W's standard heavy freight locomotive. By late 1906 the road began receiving the first of 125 of these engines assigned to Class M. The order was split between Baldwin and Alco/Richmond. In many respects, the engines were similar to the popular W2 2-8-0s. They had the same 21" x 30" cylinders on 56-inch drivers allowing the use of common parts, 200 lbs. steam pressure, and exerted the same 40,163 lbs. of tractive effort. The Class M 4-8-0s however, weighed 12-1/2 tons more than the newest and heaviest of the W2s, and had a firebox heating surface 20% greater. With the firebox located over the drivers, this weight was applied for adhesion. The 56-inch drivers allowed adequate room for the firebox. The resulting increase in power was enough to satisfy the steaming demands of heavy coal trains of the Elkhorn grade without the additional costs of roadway upgrades.

A follow-up order was placed in 1907 for 100 engines, split evenly between Alco/Richmond and Baldwin. All specifications were the same, excepting a small dome for receiving feedwater (glossary) from the injectors for the check valves and the substitution of Walschaerts valve gear creating the M1 subclass. The valve gear proved to be a less-reliable performer with rapid link block wear in the coalfields when compared to the Baker gear on the Class M 4-8-0. Otherwise, the locomotives were better steamers and faster than the W2 2-8-0s on the Elkhorn grade.

N&W was not through tinkering with the 4-8-0 wheel arrangement. In 1910 Baldwin delivered 50 twelve wheelers which ranked among the largest 4-8-0s ever constructed. Larger 24" x 30" cylinders fed steam from a 91" diameter boiler to N&W's favored 56-inch drivers. An additional 10 locomotives were constructed by the road's Roanoke Shops in 1911 and 1912. These were designated as Class M2.

The later introduction of the Class Z 2-6-6-2 brought an end to further development of the 4-8-0. The M2 4-8-0s hardly had a chance to prove there worth in the coalfields before again being overtaken by burgeoning coal traffic. Rated at 550 tons on the Elkhorn grade, the increasing coal loads resulted in double heading, making the newly perfected Mallet compound an economic alternative.

Despite prior failures with the Vauclain and Class B 2-8-0 cross-compounds, the railroad continued to show interest in compounding. The Mallet compound was introduced in 1904 in the United States when Alco delivered an 0-6-6-0 to the B&ORR. Mallet compounds soon followed from various builders to Great Northern and other railroads. In 1909 Alco/Richmond delivered the first Mallet compound to a Pocahontas road, a 2-6-6-0 to neighboring Virginian. The following year both N&W and C&O received their initial Mallet examples. In order to accommodate this new and promising type of locomotive, some reconstruction along the railroad was necessary. Longer turntables and lengthened stalls at the roundhouses from Bluefield to Norfolk, the intended area of operation, were required.

In 1910, Alco delivered five 0-8-8-0 Mallet compounds designated as Class X1. These engines were ordered for helper service and road train operation on

The Class X1 0-8-8-0 was the first successful articulated compound on N&W. The locomotives were used as helpers on the Elkhorn grade and later in heavy yard assignments.

the steep Elkhorn grade between Eckman and Bluefield. They were large, weighing in at over 188 tons. Working with 75 square feet of grate area and 200 lbs. pressure, the boiler generated steam for the 24 1/2" x 30" high-pressure cylinders that powered the rear drivers. The steam was then exhausted into the 39" x 30" cylinders powering the front set of drivers. The locomotives were given road numbers 990-994, and following break-in, were put to work.

Two things were quickly discovered. First, they were powerful, with one locomotive able to push 1050 tons on the Elkhorn, nearly doubling the tonnage of the W2 and M class locomotives. Class X1 engines 991 and 992 fell into regular use as helpers, working on the hill for many years. In this capacity they released many of the W2 2-8-0s and M 4-8-0s for road service. Secondly the railroad discovered the X1 was not a very good road locomotive. They tracked poorly at any speed, the thrusts from the low-pressure cylinders under the front engine caused a rocking motion while at higher speeds, they beat up the track and would scrape the sidewalls when passing through Elkhorn tunnel.

Baldwin delivered five 2-8-8-2 Mallet compounds in 1910, which featured many new advances the builder was promoting. N&W wished to test these features, which included a new chamber ahead of the boiler to be used as a feedwater heater, and a reheater mounted in the smokebox. Given the railroad's prior history with the Baldwin Vauclain compounds, it wanted assurances from the builder concerning the reliability of the engines. It succeeded in getting Baldwin to write a clause into the specifications which guaranteed each Mallet would replace two 4-8-0s.

As with the Mallets received from Alco, the Baldwin 2-8-8-2 engines were big, weighing 261 tons, and developing 104,300 lbs. of tractive effort. Some specifications were identical to the Alco X1 0-8-8-0s; the boiler had the same 75 square feet of grate area and worked at 200 lbs. of steam pressure. The Baldwin boiler was very long. To keep the flues to a reasonable length, some locomotives used a combustion chamber ahead of the firebox. The Baldwin 2-8-8-2 lacked this feature and instead shifted the tubes backward, leaving the front of the boiler open. In this forward section was placed a patented Vauclain "reheater" flue-type feedwater heater. The apparatus worked by transferring additional heat exhausted by the flue gases generated in the combustion process. These flue gases were channeled into this open section in front of the tube sheet but behind the smokebox. These were then directed through a series of tubes and used to preheat water before it entered the boiler. The steam was first delivered to the 24-1/2" x 32" high-pressure cylinders on the rear set of drivers, then passed it along to the low-pressure cylinders at the front set of drivers. The locomotives were classified Y1 and assigned road numbers 995-999.

In operation, the Baldwin Y1 2-8-8-2 proved to be most troublesome. They earned a poor reputation that exceeded that of the first Vauclain compounds. The improvements worked in theory, but not well when subjected to the rigors of road work. The space between the boiler flues and the feedwater heater flues was not blocked off entirely from the smokebox. As a result, it became clogged with cinders which blocked the movement of gases. The Y-1s soon had a reputation as bad steaming engines. In road trials they were frequently delayed by poor combustion. When the reheater became clogged the locomotive had to be withdrawn from service and the boiler disassembled in order to affect repair. It was evident the appliances did not work properly and further applications were dropped from the Baldwin catalog. The Y1s operated

Class Y1 engine 999 lays over in West Roanoke yard on June 8, 1910. The locomotives did not perform up to expectations, but improvements to the wheel arrangement led to long careers for later members of the Y-Class.

An Elkhorn crew stands for their photo with the 1439. The Class Z engines were the first truly successful articulated compound locomotives on the railroad. They had power and speed and quickly came into regular use in the coalfields.

east and west out of Bluefield as N&W earned a return on it's investment, but never achieved any measure of success.

N&W's two classes of Mallets gave less than hoped for results. Neighbor Chesapeake & Ohio had a 2-6-6-2 Mallet, received at the same time turning in excellent results which N&W wished to examine. Although having fewer drivers, the 2-6-6-2 was about the same size and weight of the Class Y1 2-8-8-2. The railroad arranged to borrow C&O 2-6-6-2 773 and placed it in tests against the Y1. The C&O engine out-performed the N&W engine in every aspect, leading the railroad to place orders for copies of the C&O design.

Fifteen 2-6-6-2s, classified Z1, were received from Alco/Richmond in April and May of 1912. The mallets were the first superheated freight locomotive on the road, a great advance in performance. The boiler of the Class Z was nearly equal in size to the 2-8-8-2 Y1's with 72 feet of grate area, an 82" combustion chamber and worked at 200 lbs. steam pressure. Steam was supplied to 22" x 32" high-pressure cylinders at the rear set of drivers. Exhausted steam was directed by dry pipe to 35" x 32" low-pressure cylinders located at the front drivers, and had Walschaert's valve gear. The 2-6-6-2s produced 86,250 lbs. of tractive effort when starting a train in simple cylinder operation, then switched to economical compounding with 71,875 lbs. of tractive effort. Out on the road, the Class Z locomotives could do the work of two Class M 4-8-0s. Further production of the Class M2 4-8-0 stopped. The Class Z1 was rated for 1,100 tons on the Elkhorn grade. The railroad received 25 more of the 2-6-6-2s between September and November of 1912.

Coal train operations fell into a regular routine. Westbound "Elkhorn Turns" were called at Bluefield with two of the Class Z1 mallets assigned to each hopper delivery train. The locomotives were set up tender first, running backwards, both engines on the head end of the train. At the first stop the second engine was sent around to the rear. This engine made deliveries of empties from the back of the train while the front engine held the train with its brakes. The time needed to make pickups and set outs, with the crossovers from one main to the other, was the controlling factor for movement of other trains through the coalfield. Besides coal and passenger trains, 5 freights crossed the district in

The Class Z locomotives gave N&W a powerful compound articulated type steamer that could be used as a road locomotive both in and outside the coalfield. A test train of high capacity coal gondolas works eastward between Bluefield and Roanoke in 1910.

each direction.

The assembly of eastbound coal trains was under control of the local yard masters. Eckman Yard served as the principal coal train assembly point in the Elkhorn area. An inbound hopper delivery train from Bluefield would pull into the yard with coal loads picked up and the remaining empties assigned to mines located further west. The empties would be set out to be forwarded by another mine shifter. The two inbound Class Z1 mallets off the delivery train would then turn about for duty on the eastbound coal train, with one on the head end and the other pushing. The train would be filled out to 3250 tons and a 3rd 2-6-6-2 or 0-8-8-0 mallet tacked on the rear. The yard master would obtain mainline clearance for the train which would then depart eastward. At maximum tonnage, speed up the curving 2% Elkhorn grade was 7 m.p.h. Because of the danger of suffocation, the second pusher on the rear was cut off on the fly at Coaldale, the west portal of Elkhorn tunnel. Inside the tunnel, the grade lessened to 1.4% and when combined with the forward momentum from the 3rd pusher was enough to put the train over the top of the grade. The two mallets continued eastward to the assembly yard at Flat Top, where the train filled out to 4,000 tons. The yardmaster at Flat Top Yard would obtain clearance, and the train would depart for Bluefield.

An average round trip on an Elkhorn Turn took 11 hours. There were 18 Class Z 2-6-6-2s assigned to the coal train service pool. Each pusher crew out of Eckman yard averaged 3 trips per working day. The railroad had found a reliable heavy duty locomotive, yet the coalfield remained congested and busy. The problem had evolved from that of finding a locomotive of proper power, to finding some method of unclogging operations.

An investigation led the railroad to consider several alternatives. In 1906, an article in a trade journal told of an N&W plan to replace the Elkhorn tunnel with a 4 mile long, 4 million-dollar bore. The west portal would have been between Switchback and Ennis and the east portal was to be located near Cooper on the Bluestone River. No further mention of this tunnel ever appeared as the length would have made it inoperable with existing motive power. Probably one of the more serious proposals involved the construction of a new alignment and a 2.5 mile long tunnel under Flat Top Mountain. The grade up to the bore would have been 1.4%, which eased to 0.8% inside, but it was never built, as the length would have necessitated electric haulage. The railroad by this time had placed into operation new double track from the bridge at Maybeury, up to the west portal of the tunnel at Coaldale, increasing track capacity. Adding to this improvement, another alternative involved the construction of a third main track on the hill and a twin tunnel at Coaldale alongside the existing Elkhorn bore. This proposal was dismissed as too expensive, considering costs for construction and purchase of additional locomotives needed to operate on the third track. Still, pressure remained for the railroad to find some answer to the problem.

The railroad became interested in electrification as a solution to traffic congestion in the coalfield. N&W's chief engineer, Charles Churchill, looked at the few operations underway in America. Finding nothing of merit for N&W's problems, the search went to Europe and to Italy in particular. Many different operations were underway in the Alps.

The railroad was interested in a newly developed three-phase A.C. (glossary) system which offered the advantages of great ruggedness and uniform torque. This system was particularly well adapted for regenerative braking, (glossary) an untested theory that held promise for cost savings. The very first application of a three-phase system was the Val Tellina, a dual catenary electrification which had been designed and built by the Hungarian firm, GANZ. This company built the locomotives which used induction motors that rotated at a constant speed depending on the power frequency. GANZ also built the Val Tellina railcars. Later when GANZ withdrew from the project, the Italian Government invited the American firm Westinghouse to continue the work.

During the early years of electric traction, D.C. (glossary) systems were developed more rapidly and held the advantage over A.C equipment. Beginning in the 1880's, important advances made A.C. technology more practical and efficient. Westinghouse, who had invented the automatic air brake, had entered the field in 1885. It acquired the rights for a system for distributing A.C. power from two European inventors. Shortly afterward, it constructed the first commercially successful generating station. An impressive team of electrical engineers including Nikola Tesla joined the firm. Westinghouse solved many of the problems of A.C. It developed and patented an improved transformer (glossary), based on prior designs that could step A.C. voltages up or down. Development of heavy-duty motors gave the company all that was needed to enter the railroad field. The superiority of A.C. for large scale electric power applications came to be universally recognized. Still, strong differences of opinion remained among electrical engineers concerning A.C. or D.C. power for heavy railroad traction. In general, previously developed D.C. systems tended to remain, while newly developed operations chose A.C. power.

Westinghouse gained valuable 3-phase A.C. experience when it invited the Hungarian engineer Kalman Kando to join the firm. They appointed him to the position of chief designer for the newly organized Italian Westinghouse Locomotive Company in Vado, Italy. He designed for Westinghouse the famous E55 and E33 models. The beginning of World War I forced Kando's imprisonment as a foreign national in 1914. He later escaped through Switzerland, and had an important role in Hungary's single phase electrification.

In the meantime Westinghouse's expertise continued to grow. It collaborated with Baldwin to become the leading builder of electric locomotives in North America. Advances in electric propulsion did not come easy. Controversy existed with rival General Electric boosting D.C. systems. Westinghouse proposed A.C. power with its higher transmission efficiency. They experimented with high voltage single-phase 3,300 volt A.C. current, testing a locomotive on its plant railroad at East Pittsburgh PA, in 1905. These tests led to the development of an 11,000 volt system for heavy railroad operations. Electrification offered several advantages in theory that N&W liked. One feature, long proposed, was electric braking. This utilized the energy of a moving train on a descending grade to drive the motors as generators and return power to the line. Electric, or regenerative braking, was first tried again in Italy on the Giovi line. This was a light railway were loads never exceeded 400 tons.

N&W was interested in using regenerative braking on the Elkhorn, where it could work to advantage by reducing equipment wear. The railroad decided during the summer of 1912 to electrify the Elkhorn coalfields, but the decision remained internal. System wide improvements had been underway to increase capacity to handle coal. Since reorganization on October 1, 1896 until June 30, 1915, the railroad had added 459.32 miles of new mainline and branches, 470.30 miles of second mainline track and 3.45 miles of triple track. An additional 816.11 miles of sidings had been added along single track portions of the line.

Between Bluefield and Norfolk, the rebuilt alignment and double track eased the coal transit. In 1913 at Lamberts Point in Norfolk, N&W placed into operation the largest steel coal pier on the eastern seaboard. The time for a solution to the Elkhorn problem was at hand. ■

3 A New Way of Working

Unofficial word of N&W's plans for the electrification of a 30 mile section of its mainline between Bluefield and Elkhorn was leaked on April 19, 1913, in the technical press. The June 13, 1913 edition of *Railway Age* carried the officially endorsed report, and stated that the railroad had selected Gibbs & Hill of Pennsylvania Station, New York, as consulting engineers. This firm was soon at work, and analyzed systems in use in this country and abroad, and studied local conditions as well as comparing capital and operating costs. N&W proposed to accomplish something that had never been attempted before. They intended to electrify a mountain district that carried heavy tonnage. Gibbs & Hill established the requirements using the steam operating data from the railroad as a base, which were matched with commercial manufacturer's specifications for equipment. (See map below.)

In a follow-up report dated August 8, 1913, Gibbs & Hill recommended the adoption of a three phase A.C. system, (glossary) and on August 23, 1913 the contract for electrical equipment was placed with Westinghouse Electric and Manufacturing Company of Pittsburgh PA. Specifically the system used a 25 cycle, (glossary) single phased current (glossary) of 11,000 volts supplied at the trolley wire (glossary). It was delivered as single phase and converted into three-phase power in the locomotives for traction purposes. A powerhouse (glossary) was to be erected with the initially installed capacity of 27,000 kilowatts (glossary). In order to lessen power losses in transmission, the voltage was to be raised to 33,000 volts, with the transmission lines (glossary) carried mainly on the catenary bridges (glossary) to reduce costs. Feeding points from the transmission lines (glossary) to the trolley wire were to be established at suitable locations along the line sufficiently close to minimize voltage drop (glossary) and inductive interference (glossary).

The electrical system was a radical and relatively untested departure from the D.C. and A.C. systems that had previously been used in railroad applications. N&W's proposed electrification was characterized as a heavy tonnage mountain operation. D.C systems then in service were unsuitable due to power transmission losses over long distances. Gibbs & Hill believed the constant speed three-phase induction motor, (glossary) its simplicity and inherent ruggedness, steady torque and adaptability to regenerative braking, (glossary) would be superior to both single phase A.C. systems and D.C. motors.

N&W's operating plan called for electrification of eastbound coal train service only. Coal moved in regular pre-determined fixed tonnage trains and did not suffer from fluctuations that occurred in merchandise and passenger service. Motive power needs were simplified as general freight and passenger traffic would continue to move behind steam. The Gibbs & Hill study indicated 24 electric locomotives with induction type motors could cover coal service. The control and

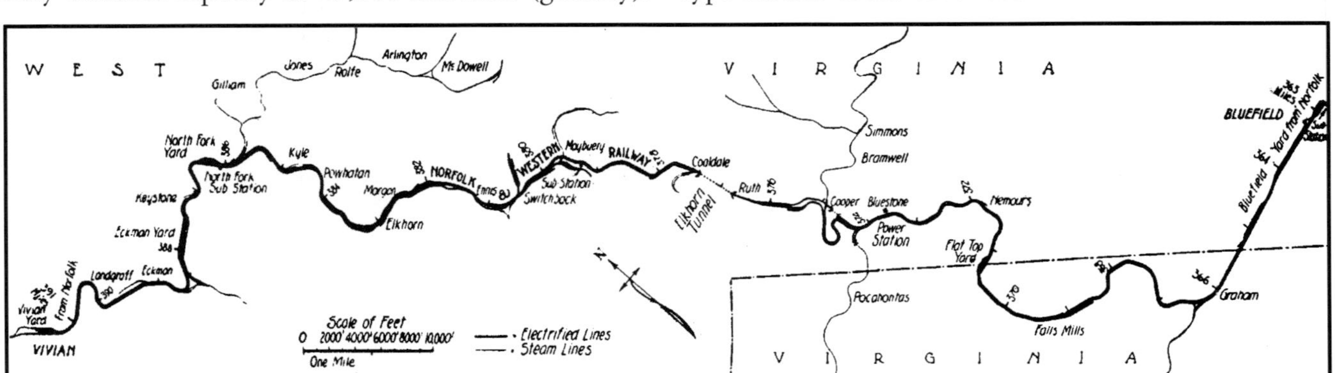

This map shows the initial territory to be covered by Norfolk & Western Railway's electrification between Bluefield and Vivian, WV. It apparently accompanied an article on the project which appeared in the June 12, 1915 issue of Railway Review.

MASON Y. COOPER COLLECTION

connections of the motors were to be arranged to give three running speeds of approximately 7, 14, and 28 miles per hour. The engines would start from a stop and accelerate up to these limits, depending on need and track restrictions. Attention in putting together the system was given to the application of regeneration and electric braking (glossary) on grades. This had never before been attempted in heavy railroad operations, but promised to decrease wear and tear on train equipment and, under favorable conditions, reduce demands on the powerhouse.

The heart of the electrification project was the power distribution system. Gibbs & Hill recommended an 11,000-volt, 25-cycle, single-phase A.C. system similar to those the firm had installed in earlier projects involving electrification on the New Haven and the Boston & Maine Railroads. Prior three-phase power distribution systems in Europe featured a complex two-wire overhead catenary, which single phase avoided. Greater efficiency in power transmission was gained from a single phase which was believed to be of importance with the heavy requirements of N&W's system.

Westinghouse, along with rival General Electric, had experimented for years with some method by which single phase power could be split to be used in three phase induction motors. General Electric had succeeded in constructing a small experimental unit with this feature, but the concept remained untried by any railroad. N&W's plan to use the split phase system for one of the largest electrification projects ever attempted was a calculated gamble. If it succeeded, it would be an important step forward for the rail industry.

The site selected for the power station was Bluestone Junction, eleven miles west of Bluefield, WV. Water requirements for the boiler (glossary) and condensation (glossary) purposes determined the location, as Bluestone was practically the only point where enough water existed for the plant. The Bluestone River arose in a region immediately to the south, and flowed in a northerly direction to empty into the New River. Power requirements called for generation of sufficient power to handle 65,000 gross tons in a 24 hour period. Specifically, the plant had to generate 8,000

The site selected for the power station was Bluestone Junction, where water requirements for the boilers could be met. The electric locomotive repair shops were located here also.

HP per train for startup and maintain 6,000 HP per train for operation. Before final site selection was made, commercial alternatives to plant construction were explored. These were rejected when negotiations failed to bring utility electric rate charges in line with power production costs that had been previously determined by N&W.

Construction at Bluestone Junction began in the spring of 1914. The main powerhouse building was built by first assembling a steel frame on a concrete foundation. The walls were laid with hand-burnt brick, described in a brochure as "plain and substantial in character." The finished structure was 135 ft. wide by 158 ft. long, with a room extension measuring 52 ft. by 33 ft., located on the long wall in the northeast corner. The main building was divided internally into two rooms by a center wall running the entire length of the structure, creating a 79 ft. 3 in. by 158 ft. boiler room and a 56 ft. by 158 ft. turbine room. Across the east end from the turbine room, a 26 ft. wide section was subdivided and assigned for switching equipment, offices and other plant facilities. This room was fitted with intermediate floors and galleries. On the ground floor of the extension to the building were located the step-up (glossary) and step-down (glossary) transformers (glossary). On the second or mezzanine floor of this extension section were the locker and toilet rooms for the electric locomotive engineers and firemen. Also on the second floor was the high tension switching equipment (glossary). Above this level, which opened onto the main floor of the powerhouse was the low tension switching room (glossary). This extension section was open to but separated from the turbine room by heavy wire screens. The next floor up was the operating gallery, which overlooked the turbine room. This level contained a small office and lavatory. The top floor was used for offices.

In the basement level of the powerhouse were located the pumps, oil filters, air compressors, fly ash handling equipment and forced draft fans. This basement floor was located at ground level. Along one side of the basement and in a separate room was found the main plant material storeroom. Replacement parts to be used in the powerhouse were stored in this area. The main floor contained the boiler room, located at a level 13 ft. 6 in. directly above the basement. Here were located the boilers, and the accompanying stokers (glossary), stoker engines (glossary), and feedwater heaters (glossary).

Water for all plant purposes except drinking was taken from an intake canal running from the Bluestone River. Two rows of 5 ft. x 7 ft. screens constructed of steel frames were located at the intake to treat river water. Three sizes of decreasing sized wire mesh were fitted onto the screens to remove particulate (glossary). Water was drawn from a cooling pond (glossary) during certain seasons if the supply of river water was low or not suitable for boiler use on account of mud. Recirculating water from the condenser (glossary) pumps was discharged into cooling ponds and from there discharged again into the intake canal (glossary). The cooling pond was located next to the powerhouse and was 90 ft. by 296 ft. and 5 ft. deep. The pond was equipped with 280 nozzles, with space provided for 140 more, and capable of cooling 16,800 gallons per minute. Water was sprayed from nozzles attached to the outlets from the condenser pumps and heat dissipated from this water into the atmosphere. As long as sufficient cold river water of suitable quality was available, the water from the condenser pumps was not sprayed into the cooling pond, but allowed to flow directly into a river reservoir located several hundred feet below the intake. Heat contained in the water dissipated into the atmosphere from this reservoir, albeit at a slower rate. It then circulated back via a connection to the intake, a circuit of about 1400 ft.

Powerhouse coal was delivered by hopper cars on a railroad siding installed on the south side of the facility, adjacent the N&W mainline. Cars were spotted for discharge into a hopper bin located underneath the tracks. Below this bin was a single roll crusher which reduced the coal to a smaller and more uniform size.

The crusher emptied into a continuous bucket inclined conveyor. Conveyor capacity was 60 tons per hour when operating at a speed of 90 feet per minute. The powerhouse conveyor was similar to other installations working at several mines in the region, benefitting from common parts should emergency repair become necessary. Electric A.C. motors drove all coal handling machinery. The conveyor lifted coal to a storage hopper located in the eastern portion of the boiler room. From a storage hopper located here, the coal was fed into a horizontal distributing conveyor into two other storage bins, which were suspended above and between the boilers. These were attached to the building girders, with each bin having the capacity of 350 tons. Coal was fed by gravity from the storage bins into the boilers. There were a total of ten 2000 HP Stirling-type water tube boilers (glossary), arranged in two rows, facing each other with a firing aisle located between. Each row consisted of two batteries of two boilers each and one single boiler. Space was provided in the powerhouse for the later installation of four more boilers, should expansion become necessary. The boilers were designed for operation at a maximum 225-lbs. steam pressure. Each unit had a total effective steam heating surface of 6,772 sq. ft. which was comprised of 3 steel 42 in. x 16 ft. 4 in. drums and 487 fire tubes. The boilers were also equipped with superheaters capable of raising steam another 150 degrees F. at normal rating.

Each boiler was equipped with an underfeed stoker with a capacity to evaporate 61,000 lbs. of water per hour into steam, at 200 lb. pressure. The feedwater heater pre-heated the water to 200 degrees F. which was then evaporated in the boiler before the superheaters raised the temperature the additional 150 degrees F. The boiler stokers were capable of operating at three times the normal boiler rating when burning coal rated at 12,250 B.T.U (glossary) per pound.

Water was taken from the Bluestone River for use in the powerhouse, then discharged into cooling ponds (located in the foreground) and made available for re-use in the plant.

ABOVE: Coal for the powerhouse was received from a siding located on the south side of the power plant. The cars discharged into a storage hopper located beneath the track.

BELOW: Incoming coal for the powerhouse was weighed in this scale house located at Bluestone Junction. A storehouse is located adjacent to the scales.

A store house and supply house was located at Bluestone Junction adjacent to the N&W mainline where it was used to supplement operations at the facility.

A forced air draft installation consisted of three Sturtevant E-5 type multivane turbine driven fans located in the boiler room basement. These were driven by the steam turbines through a 4.1 herringbone reduction gearing (glossary). Each Sturtevant turbine had a normal rating of 300 HP. The smokestack was located next to the boiler room and was made of radial-type brick (glossary), and was 268 feet tall. The boilers and all accessory apparatus existed to provide steam for the adjoining turbine room.

Located on the main floor of the turbine room were the main turbo-generator units, exciters and signal generators. This room was located 18 ft. above the basement, slightly higher than the boiler room found across the center dividing wall. A 30-ton capacity gantry crane spanned the turbine room. The condenser equipment and accompanying fans for cooling the generators were located directly below the turbine room in the basement. The pumps and filters were located off the main area in the room extension.

Initially, the powerhouse was equipped with three generating units, with space allowed for a fourth. The generators were Westinghouse-Parsons impulse reaction double flow-type horizontal turbines. Each was rated at 10,000 kilowatts when operating at an 80% power factor, 11,000 volts, 25 cycle, single phase A.C. This rating was reached when operating at 190 lbs. steam pressure supplied by the boiler room.

An oil relay governor operated the turbine steam valves. The oil was maintained at an operating pressure of 45 lbs. for the pressure mechanism. This same oil was reduced to about 10 lbs. pressure and was used for lubrication purposes. If the oil supply was interrupted, the safety stop valve automatically operated to shut off steam supply, preventing damage to the generators. A speed governor tripped the automatic stop valve whenever the speed of the turbine exceeded a predetermined value. This value could be adjusted by the operating staff.

Normal ratings for 24 hour generator operation specified a rise in temperature not to exceed 60 degrees C. above the cooling air. The generators could deliver a single phase output of 12,500 kilowatts at 80% power factor for 5 minutes without injury from overheating. The two-pole field coils were wound with copper strap, and insulated for one minute potential of 1,000 volts. Provision was made on the generators for plugs to be utilized at some future date for balancing. (Balancing was a method of reducing electrical induction interference from the catenary on nearby telephone and power lines by running an adjacent wire of opposite polarity. The presence of a balancing wire in a catenary system canceled out the interference, but significantly increased wire expense). Little extra cost was added by

N&W specifying plug-equipped field coils so that at some point in the future a balancing wire could be added to the system which utilized taps from the plugs of the existing field coils.

Generator armatures were wound for three-phase power and insulated to endure a one minute voltage test of 22,000 volts without damage. The ceramic insulation could withstand a temperature of 90 degrees C. without injury. A continuous temperature record was furnished to the operator from thermo-couple sensors imbedded in the armatures. A No. 6 Sirocco blower with a capacity of moving 50,000 cu. ft. of air per minute ventilated each generator. Each turbine generator was equipped with a separate Le Blance Jet-type condenser. The condenser injector, water and air pumps were driven by a horizontal turbine. The air

The powerhouse generators were located in a separate room at Bluestone. Three Westinghouse-Parsons impulse type generators were initially installed with space provided for a fourth unit.

pump discharged into the intake canal while the ejection pump discharged used boiler water into a pipe leading to the spray cooling pond.

There were two turbine-driven and one motor-driven exciter generators. The two turbine driven generators were small Westinghouse 150 kilowatt horizontal turbines. These rotary armature type generators with commutating poles were compound wound, each having a rating of 600 amperes at 250 volts when running at 2,200 r.p.m.. The motor-driven generator consisted of a 3-phase induction motor rated at 150 kilowatts at 440 volts, 25 cycles when running at 720 r.p.m.

The six-pole compound wound generators had a normal rating of 600 amperes at 250 volts, receiving power from a 3-phase self-cooled oil transformer operating at 200 kilowatts A.C. A Tirrill regulator was used to vary the exciter voltage and current over a sufficient range to maintain a constant voltage on all three main generators, under all anticipated operating conditions.

A monitoring board for the generator operating crew was installed on the south wall of the turbine room. Symbols governing the operation of the main generators, exciters and signal generators were painted on ground glass and illuminated by the push button switches in the operating gallery. Return signal lamps were also provided on the main power board upstairs in the operating gallery. Pneumatic whistles were provided in high noise areas, the boiler room, pump room, and turbine room. Additionally, a pneumatic whistle was installed at each boiler as part of the signal communication system.

The line side signal system in the Elkhorn coalfield segment, prior to electrification, was equipped with D.C. powered automatic signals. These had only recently been installed, and were mounted on line side poles that required primary storage batteries for operation. The signals in the new electrified zone were converted to A.C. power. The D.C signals, being practically new, were taken down to be used on other parts of the railroad. The A.C. powered signals took advantage of power provided by the electric train installation. Signal system power was supplied by an auxiliary steam line which drove two turbine driven 50 kilowatt 4,400

Each of the main generators could produce 10,000 kilowatts when operating at 80% capacity, and could run for 24 hours without a significant rise in temperature.

volt, 60 cycle single-phase generators, installed for that function.

Power for electric train operation was supplied from the three primary turbine generators, which were wound for 11,000 volts, 25 cycle three-phase A.C., delivered to a power bus. The switching apparatus sectionalized the power bus, and auxiliary power was taken from one of the sections. The bus was constructed of copper tubing instead of wire, and carried on ceramic insulators mounted on pipe framework. Copper tubing and bare wire were used in the power bus wherever possible, with insulated wire used only where the conductor was carried in a conduit.

The power was stepped up from 11,000 volts by first passing through three 5,000 k.v.a. step-up transformers to 44,000 volts. The step-up transformers were controlled by three 2-pole type oil circuit breakers. The high tension power then passed upward through openings in the ceiling to the next floor where the oil circuit breakers and the 44,000 volt power bus were located. Again, copper tubing and bare wire were used exclusively in this bus work and wiring. The 44,000 volt bus was sectionalized with two high tension feeders taken from each side. The high tension feeders and trolley feeders then left the building by openings located in the south wall. Lightening arrestors located on the roof grounded the powerhouse. The operating gallery controlled the oil circuit breakers on the power bus. The high voltages necessitated the use of knife-style disconnecting switches between the busses and all oil-type circuit breakers. The high-tension line and trolley-wire circuit breaker feeds also had knife-style disconnecting switches.

As a safety measure, the electrical switching and control equipment in the operating gallery received its power from the auxiliary D.C. bus or a storage battery located in the northeast corner of the turbine room basement. The storage battery had a capacity of 160 ampere hours at 220 volts, and was charged by a 5 k.w. motor generator set located in the operating gallery. The switchboards in the operating gallery consisted of a generator bench, main power board, auxiliary power board, and a signal power board. These were under the control of the Power Director.

A novel part of Norfolk & Western's electrification installation were the regeneration loading rheostats. In normal train operations the trolley wire returned excess regenerated power to the power house. The excess passed to the 11,000 volt bus and through the various transformers back to the generators. This in theory would reverse the generators and operate them as motors. To prevent this from happening, a loading device was incorporated into the circuit that consisted of electrodes immersed in the intake canal and controlled by switches. An automatic switch controlling the loading device was composed of a group of relays, magnetic switches, and current transformers connected in a circuit. When the amount of excess regenerated power from a train moving downgrade reached a certain pre-set value, a relay closed throwing one water rheostat onto the 11,000 volt bus connected to the trolley wire. If the amount of regenerated power thrown off by the train exceeded the capacity of one water rheostat by 300 k.v.a., another relay closed, throwing a second water rheostat into the 11,000 volt trolley circuit. The difference between the amount of excess power and the capacity of the rheostats in the circuit was made up by the main generators.

With two water rheostats in service, when the excess regenerated power from a train moving downgrade dropped, then one relay opened the breaker that closed first, which then cut the rheostat out of the circuit. The other water rheostat remained in service until the excess regenerated power dropped. When the excess regenerated power from the moving train reduced to zero with one rheostat in the circuit, then one of the tripping relays would open the circuit breakers, which dropped the last rheostat off the 11,000 volt bus. The water rheostats consisted of a steel cone and were located outside the transformer house at the power station. This cone was connected to a lead wire from the circuit breakers, and to a fixed ingot iron plate located at the bottom of the intake canal. The fixed iron ingot was grounded to a copper plate embedded in the earth outside the canal. By changing the distance between the steel cone and the iron plate in the bottom of the canal, the capacity of the rheostat was adjusted. The cone was raised or lowered in the water of the canal by means of a hand-operated winch carried on a steel cable from across the intake canal.

High-tension power generated at the station was transmitted as 44,000 volts, 25 cycles, single phase A.C. by feeders leaving the building by the openings built into the powerhouse south wall. The high-tension wires were carried on the catenary support all along the railroad, except in tunnels. At these points, the line went above ground, away from the right of way and over the summit. There were two single-phase lines between the powerhouse and the substations. One high-tension line was used for power transmission, and the second served as a backup for maintenance. These power lines were each constructed of four strands of No.2/0 seven-strand hard drawn copper wire. Also carried along with these lines were a 3/8-in. diameter steel wire which served as a ground.

The insulators on the high-tension power line were four bell, or petticoat pin type, with a maximum diameter of 12 inches. They were tested at voltage ratings of

High-tension power left the powerhouse at 44,000 volts, 25 cycles and was carried along by an arm located on the left side of the catenary superstructure.

165,000 and 120,000 volts respectively, for dry and wet weather condition flashover. The ultimate mechanical strength of the insulator was about 4,000 pounds when applied to the pin at the wire groove, where the stress of the power line was carried.

The wooden pole line, which crossed tunnels and summits, was made of chestnut, and varied from 45 ft. to 55 ft. At the time of installation, the pole line carried three arms, with provision for a fourth. The two upper arms carried the 44,000 volt wires and the lower arm the 4,400 volt signal wires. An 11,000-volt trolley feeder was also carried over Elkhorn tunnel on the wooden poles as a safety precaution to eliminate the chance of the section between the Maybuery substation and Coaldale becoming isolated in the event of a derailment in the tunnel. The cross arms on all the poles were connected to the ground wire which terminated in a copper plate imbedded in the earth at the base of every fourth pole.

The high-tension pole line delivered power to five line side substations. These substations housed step-down transformers, which reduced the 44,000 volts high tension power to 11,000 volts for train operation. These step down transformers were Westinghouse brand single-phase, oil-insulated, water-cooled transformers and equipped with thermostats. Should high temperatures from excessive operation occur, the thermostat would close and an alarm bell circuit activated in the nearest operator's office. A transformer oil filtration outfit was supplied to N&W with the installation and transferred from one substation to another by maintenance forces on a rotating basis.

Substations varied in capacity according to expected demand. This capacity was determined by steam operating data supplied to Gibbs & Hill from the railroad. Each installation was equipped with oil circuit breakers as a safety measure. The remote control circuit breakers were operated from nearby signal towers, passenger stations or yard offices, and required no extra attendant. Each line side substation formed a section of the high tension power distribution line. In this manner, any difficulties that occurred were localized, allowing the rest of the system to remain in operation. This was accomplished by means of air break switches on the substation roof.

The first substation was located at Bluefield, WV., the eastern limit of the electrification. The Bluefield substation housed two 3,000 k.v.a. Westinghouse transformers that were controlled by the Bluefield yard master. The second substation moving westward was located in the transformer and switching house at the Bluestone power station. Due to the potential damage from a short circuit, it was decided not to feed the nearby sections of the trolley wire directly from the power station bus. A mishap had the potential to take the powerhouse off-line, shutting down the whole system if power was taken directly off the bus. The substation at Maybeury, which was located on the steepest part of the Elkhorn grade, had the greatest capacity. This supplied power where the greatest demand was anticipated. The Maybeury substation housed two 5,000 k.v.a. transformers and was operated from the Maybeury passenger station. The substation at North Fork was controlled from the signal tower and featured two 3,0000 k.v.a. transformers. The last substation was at Vivian, and was equipped with one 2,000 k.v.a. Westinghouse transformer, and controlled by the yard master.

One advantage gained by using the high voltage on the contact system was that no intermediate feeders were necessary. Distribution of power was made directly from the substations to the trolley wire.

The substations supplied 11,000 volts A.C. to the catenary for train operation. Restrictions placed by

Mainline catenary installation, as seen here at Bluestone, WV, consisted of two tubular steel poles supporting an H-beam, to which all other accessories were then attached.

mining tipples, buildings, and coke ovens hindered catenary construction along the right of way. Another difficulty was that 60% of the line was on curves ranging up to 12 degrees, which posed other problems with catenary design. Two distinctive types of structures were used for supporting the catenary system. One design was used on the main line, and was distinguished by having the wires suspended from overhead steel beams and girders. The second design was used in yards. This differed in appearance as a long overhead cross-catenary cable was used to support longitudinal catenary wires over a large number of tracks.

The typical mainline installation consisted of two tubular steel poles set in concrete foundations. Foundations were constructed of 1:2:5 concrete (glossary), reinforced with rods and cored in a collapsible wooden form. A copper ground plate was set in coke underneath one of the two towers and a large copper cable was carried up through the concrete for later attachment to the pole for electrical grounding. The tubular steel poles were built up of various lengths, sizes and weights, wedged, then welded together to form a single solid unit. They carried between them a structural H-beam onto which the bulk of the catenary was carried. Cross beam sag braces were installed on each of the two tubular steel poles to help support the H-beam. The tubular steel tower poles extended above the H-beams to support cross arms. The high tension feeder and signal lines from the power house were attached to these support cross arms. Ground wires were clamped to castings attached to the top of the pole and connected to the copper cable carried up through the pole. The catenary structure was normally unsupported by guy wires if installed on a tangent section of track. On curved track, the catenary structure was guyed against the curve.

Normal catenary tower installations were set 300 ft. apart on straight track and somewhat closer together on curves. Exact placement on curves depended on the degree of curvature and whether intermediate pull-off poles and guys were to be used between adjacent supporting structures. The guys were of substantial construction as they were intended to carry a portion of the load. The guy anchors were made of concrete slabs containing two or more pieces of old rail for added weight. When double guys were used, the two rods were set at an angle of 20 degrees with the pole. Steel structures used in the catenary, other than the main tubular poles, were all fabricated in Roanoke by the Virginia Bridge Company.

Variations included catenary bridges that had both structural posts and girders. These were used wherever the structures carried both line side train signals and the catenary system. Similar but heavier structural steel bridges were used on curved track where property for installing guy wires was not available. These factors contributed to making almost every bridge a unique installation. In the yards, the catenary structure consisted of two or more high strength steel posts that were either guyed or self-supporting depending on location. From the top of one post, a cross-catenary cable of extra high-tension galvanized steel stranded wire was run across the tracks to the top of the opposite post. This wire replaced the H-beam used in mainline installations. In many instances, this was a 1-1/2-inch steel cable that had tensile strength of 240,000 lbs. It was designed to have a large sag which, where the yard was several tracks wide, required high posts. A second strand of steel cable was carried horizontally below the first strand and suspended from the top of the cross-catenary cable with 3/4 in. rods.

Seemingly complex, the catenary system was actually a simple design which avoided many problems experienced in earlier systems. Its primary objective was to hold the 3/0-gauge trolley wire at a uniform height of 24 ft. above the top of the rail and not vary in location more than 12 inches from track center. Allowances were made for super elevation of curves. The end result was that the bottom of the contact wire presented an unbroken level surface on which the locomotive pantograph rode.

On tangent sections of main line track the construction was simple. From each of the supporting structures a steel casting was attached to the horizontal beam serving as a support for three suspension insulators. Each of these three insulators were tested for a working voltage of 11,000 volts without flashover. They were arranged so that they worked together as special backup against line troubles. To the bottom insulator was attached an iron casting which supported and clamped to the messenger wire. This casting had room for the wire and arcing horns so that should an insulator fail, the resulting electric arc would most likely be from the casting rather than the wire, thereby avoiding damage to the catenary. This iron casting could be more easily changed out than a burnout on the catenary wire.

On the mainline, the 11,000 volt A.C. was relayed from the high-tension line to the messenger wire. This messenger wire was made of 1/2 in. extra high tension galvanized steel stranded cable having a tensile strength of over 30,000 lbs. The messenger wire was aligned above the auxiliary and trolley wires. The nominal sag allowed in the messenger wire was no more than 5 ft. over the maximum 300 ft. span between supporting towers. The actual amount of sag varied depending on the individual span. Spaced at intervals of 30 ft. apart on the messenger wire were bent strap clips which held hanger straps. These straps relayed the voltage downward. A pair of iron trolley clips on the bottom of these

Catenary towers were usually placed 300 feet apart. This scene looks westward along the mainline at Landgraff, WV.

hanger straps gripped an auxiliary messenger wire.

The auxiliary messenger wire was carried directly above and attached to the trolley wire. This auxiliary wire was a size 0 round copper or steel wire. Copper was used over steel depending on load conditions, when added carrying capacity was required. From the auxiliary messenger wire the size 000 grooved contact or "trolley" wire was supported by a pair of clips bolted to the auxiliary messenger and spaced 15 ft. apart. The spacing was adjusted so that they were equidistant from each strap hanger holding up the auxiliary to the messenger wire above.

The contact wire was held steady in place crosswise to the tracks on the longer tangent stretches by another wire called a "steady strand". All steady strands were insulated from an electrical short by a string of three suspension-type insulators. Sometimes the placement of a steady strand aligned with the steel supporting structure. In these examples, where one of these cross steady strands ran from pole to pole and was attached to the wires over two or more tracks, it was sectionalized between the tracks by the use of a 4 ft. impregnated wooden stick insulator.

The contact or trolley wire was sectionalized longitudinally at the substations by breaking the messenger wire with two strings in parallel consisting of three insulators each. Directly under the point of the electrical break in the messenger wire, the auxiliary messenger and contact wires were attached with turnbuckles. A heavy impregnated wood stick insulator was attached to the ends of these turnbuckles. Tee iron runners were supported both from the trolley wires and from the wooden stick insulators. The tee runners had ample electrical clearance from each other to prevent short circuiting.

Besides substations, section breaks were provided in the system to allow for maintenance through parts of the catenary which were normally connected electrically. Maintenance workers accomplished this with specially designed disconnecting knife switches that mounted on the top of wooden break insulators at specific locations. These could be opened or closed from the ground by means of a long impregnated wooden switch stick.

The catenary system was anchored longitudinally at certain locations. Dead-ending (glossary) with solidly attached cross girders of steel accomplished this. Specially constructed structural bridges were provided solely for this purpose. A double string of three insulators isolated the messenger wire from the dead-ending structural bridge. These anchors limited the zone of trouble should a wreck occur or some unforeseen incident break a messenger wire. The normal arrangement of this messenger wire support at the bottom of the string of insulators was designed so that at a certain pull, the messenger would slip through. Should a messenger wire break between anchors, it would slip through several supports adjacent to the break, distributing the abnormal strain of the break over several bridges. Otherwise, the strain would be applied to the two bridges on either side of the break and result in even greater damage.

On the largest part of the Elkhorn line, the track was curved. One obstacle occurring on curved track was the necessity of keeping the trolley wire within center-line standards. This contact wire was designed to be no more than 6 to 12 in. off the centerline of track in order to maintain contact with the locomotive pantograph. This allowed the contact wire curvature to be less than the track curvature, meaning the trolley usually struck on the inside part of the locomotive pantograph on curves.

The hangers on curves were placed 15 ft. apart and the clips at the bottom held both the auxiliary messenger and the trolley wire. This was essential to maintain a proper alignment, since the contact wire must be held down vertically below the auxiliary messenger wire and resist bending the clips and hangers. Otherwise, the curve pull of the contact wire would raise up to the same level as the auxiliary messenger wire by the added stress of the curve. Holding of both wires by the same clips in curves could, but did not result in "hard spots," (points of high stress in the contact wire) because additional support points in the form of pull-off poles were used. One and sometimes two pull-off poles were used between structural bridges.

In yards, the catenary system was similar to the mainline except that the messenger insulators were supported from the lower, or horizontal wire strand, of the cross-catenary bridge. In yards, only two suspension insulators were used between live wires and grounded structures because of the reduced incidence of wear resulting from lowered train speeds. The auxiliary messenger wire was omitted in the yards as well as the clips on the end of the hanger straps holding the trolley wire.

Inside Elkhorn Tunnel, special catenary construction was required. The walls of the 3000-ft. bore were an egg-shaped elliptical, spaced 14-ft. apart at the trolley wire level and 13 ft apart at the base. The trolley wire level was 17-ft. 1 in. above the top of the rail inside the tunnel, while the normal height was 23 ft. 9 in. for outside installations. (This demonstrates the cramped conditions that existed inside the bore, not to mention a potential source of damage to the pantograph). The top of the arch inside Elkhorn Tunnel was 19-ft. 3 in. above the rail. The general arrangement adopted for catenary inside the tunnel was to have a supporting structure located every 75 feet. On each side, two cast iron brackets were clamped to the brick lining of the tunnel by large bronze expansion bolts. By this method, a 44,000 volt transmission line insulator was carried. A 2-in. brass pipe was carried parallel to the tracks and attached by the pair of insulators, one on each side of the pipe. Another line insulator on each side was mounted by bronze castings at the middle of the brass pipe. A cross pipe was bent to follow the arc of the tunnel roof. This ran from the support pipes on each side and formed the support on the top of the arch for a bronze messenger wire. The live catenary parts had double 44,000 volt insulation, due to the insulators on each side protecting from a ground. As a precaution and due to the very limited clearance inside the bore, the installation was tested to three times normal trolley voltage.

On branches and coal mine sidings, where train movements were made at reduced speed, a less expensive wooden pole bracket type of construction was used. In these places of low frequency use, only two suspension insulators were necessary. The messenger wire installed was 3/8 in. steel strand and the contact wire a size No. 2/0; no auxiliary messenger wire was used. The overhead contact wire at mine tracks was normally kept de-energized where the danger of electrocution to workers moving about on the top of cars was a possibility. These sidings were designated as "Dead Sections" and identified by signs. A sectionalizing switch isolated this track from the rest of the system and was kept normally in the open position. This would be closed before equipment entered to work the mine. Before entering, train crews had to observe that the wire was free of obstructions and material, and all persons near the track were warned the wire was energized. When it was necessary to de-energize the overhead contact wire while a locomotive was still on the

Catenary installations in yards varied from mainline structures as can be seen in this view taken at Bluefield. Yard tracks were powered from a trolley hung from a cable extending over several tracks.

Catenary installation in the coal assembly yard at Eckman, WV clearly shows the trolley hung from a cable using wooden poles. Wooden poles also replaced the use of steel on some sidetracks.

mine track, the motor was required to drop it's pantograph as a safety precaution. The pantograph would be kept down until the wire was again energized.

Headquarters for the maintenance of the overhead catenary trolley and transmission lines was Bluestone Junction. The line repairmen were available and subject to be called any time during the day or night. On a siding located at Bluestone Junction, a repair train consisting of a tower car, tool car and flat car was held for emergencies. One maintenance man rode the revenue trains continuously to observe the catenary for defects. He would throw off notes as necessary to catenary repairmen. For ordinary repairs, small gangs of three to five men were sent out on small gasoline powered cars. Wherever possible, repairs were made from the overhead bridges or from specially insulated ladders in order to avoid interruptions in train service. Specially treated and impregnated ladders were provided to the gangs and used for light repairs to trolley fingers, clips and other details on the 11,000 volt catenary.

Certain line repairmen were stationed at one or two points on the line from which they quickly responded to problems with the help of the gasoline motor cars. A patrol telephone line was installed over the entire length of the electrified zone as an additional communication tool for inspection and maintenance. The telephone was connected to the powerhouse telephone operator and could be used to contact the train dispatcher and other officials.

The electrical circuit had to be complete to function. It was necessary to get the current from the locomotive back to the substation. This was accomplished by bonded track rails. A special track current bond which was quicker to install was designed for the electric installation. By manufacturing the track current bonds with one of the terminals attached, and crimping or looping the electric cable near one end, the bond could be installed by removing only one bolt in the fish plate. The bonding cable consisted of 37 strands of 0.0534-0.475-in. copper wire. At the sub-stations, the track rails were connected to the bus by copper cables laid underground.

At signal bridges, the rail joints were insulated in accordance with requirements of the signal circuits. Here it was necessary to install bonds so that the flow of propulsion current could be continuous. This result was accomplished by the use of impedance bonds which prevented the flow of 60-cycle signal current from getting past the insulated joint, but at the same time did not interfere with the flow of 25 cycle propulsion current.

It was necessary to protect telephone and telegraph lines along the right of way from service interruption. The primary danger was physical contact between falling telegraph wires and high tension power wires. The secondary problem was induction interference that would set up in parallel telephone and telegraph lines due to current in the power circuits. It was found best to move telephone and telegraph lines entirely to the north side of the tracks and locate them at the extreme edge of the right-of-way. In this way they were as far as possible from the power circuits.

To guard against induction interference, it was decided to install the transformer sub-stations somewhat closer together than would be required on account of voltage losses. The sub-stations were placed near the ends of the electric lines, which resulted in feeding the section of contact wire from both ends. This localized the induced voltage in parallel telephone and telegraph lines. Track transformers or track boosters in the trolley and track circuits were placed at intervals of about one mile. These provided for emergencies when one substation needed to be out of service, or high currents occurring in cases of short circuits or grounds. All of these features together formed the catenary system.

The catenary delivered electricity to the pantograph for the locomotives. The steep Elkhorn grade placed special conditions on the design requirements for motive power. N&W's prior experience with steam operations had found it necessary to have part of the motive power on the rear when handling coal trains on the grade. The most important technical feature of the electrification was the pioneering use of split phase locomotives. A three-phase induction motor without commutators used instead a liquid rheostat giving it more rugged construction traits. A liquid rheostat, besides giving the smoothest possible rate of increase in tractive effort, was able to dissipate the large amount of heat generated from heavy operating conditions. In addition, the characteristics of an induction motor allowed the locomotive on the rear to be used to hold a train at constant speed on descending grades by regenerating and returning power to the line.

Taking into account the data supplied by N&W, Gibbs & Hill determined the traffic on the electrified Elkhorn region could be handled by a dozen 270-ton Baldwin-Westinghouse locomotives, semi-permanently coupled together. The locomotive design selected had a 2-4-4-2, (1-B+B-1) wheel arrangement, with two trucks connected by a joint hinge. The box cab compartment was supported by spring cushioned friction plates and rested entirely on the main truck with no weight on the center pin. An engineer's operating compartment was provided at one end of each unit. The two units were coupled to provide for operation from either end, with an overall locomotive length of 105-feet 8-inches. The motors were the largest and heaviest electric locomotives in the United States when they were first placed in operation.

Baldwin Locomotive Works and the Westinghouse Electric and Manufacturing Company had collaborated on electric locomotives since 1896, when two initial units were first constructed for a mining operation. By prior arrangement Baldwin would construct the locomotive shells at its Philadelphia plant and ship them to Pittsburgh where they would receive their electrical equipment at the Westinghouse plant. At the time of the N&W order, a large order for single phase electric locomotives was being completed for the New York, New Haven & Hartford RR. This order featured various types to be used in passenger, freight and switching service, although none were similar in design, size or intended operation to the N&W engines.

The N&W locomotives were each equipped with eight 3-phase induction type traction motors, mounted on the upper side of the frame, with secondary windings for 4 pole and 8 pole operation. The motors were force air-cooled by means of a main ventilating duct that was built into the floor of the locomotive, which delivered air to the phase converter and to cooling towers for the liquid rheostats. There were two running speeds available: 14 and 28 m.p.h. depending on the need for torque or speed. In order to start the locomotive, an electrical resistance was inserted in the secondary circuit of the motor by means of the liquid rheostat. For the 14 m.p.h. speed, all motors were con-

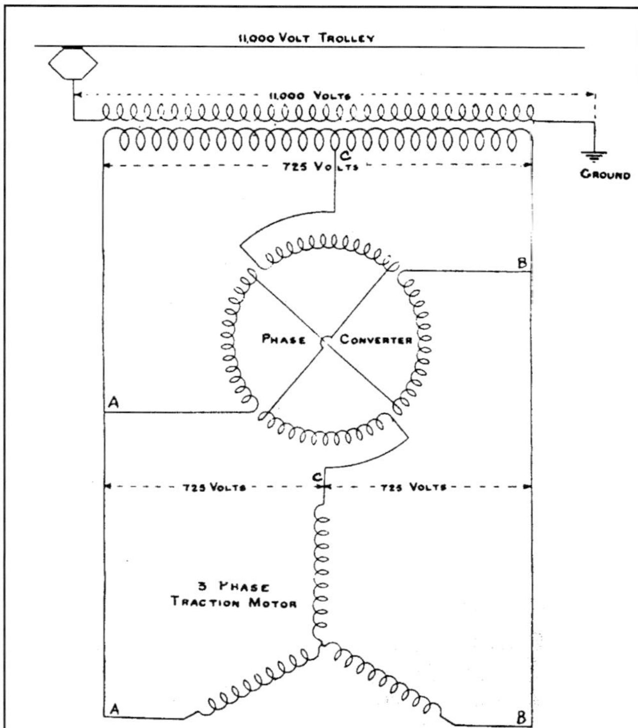

Diagram of Circuits on Norfolk & Western Electric Locomotives, From 11,000 Volts, Single Phase, on Trolley Wire, Through Step-Down Transformer, Phase Converter and Three-Phase Motor.

The circuitry of a N&W electric locomotive was both simple and rugged. Regenerative braking was accomplished by reversing connections with the phase-converter.

MASON Y. COOPER COLLECTION
ORIGINALLY FROM RAILWAY REVIEW

Baldwin-Westinghouse Class C 2500 produced 3200 HP and was the first electric locomotive delivered to N&W. This view, taken in 1921, shows the engine laying over at Bluestone between assignments.

Class C 2511, built in 1915, was the last of its class for N&W's electrification. The 2-4-4-2 + 2-4-4-2 wheel arrangement is clearly seen in this photo taken at Bluestone Junction.

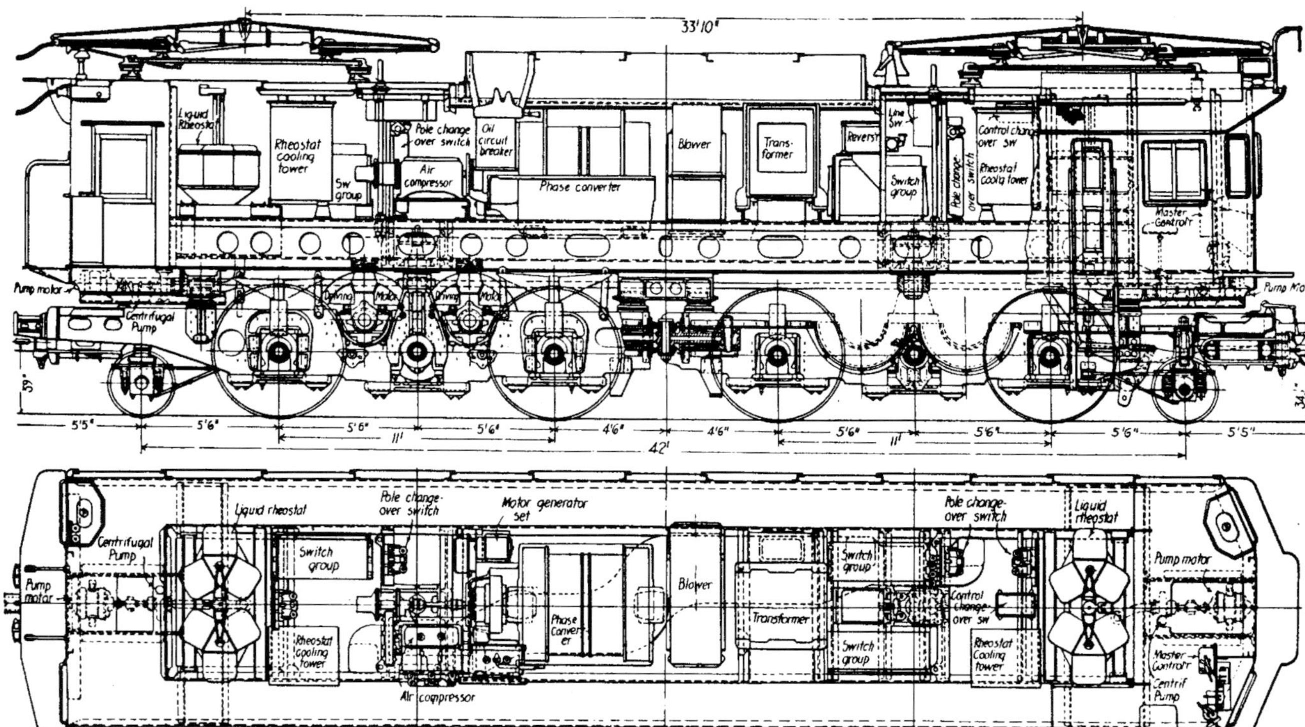

This longitudinal cross section and floor plan of a Norfolk & Western Railway Class C Electric Locomotive shows the layout of equipment inside.

Mason Y. Cooper Collection
Originally from Railway Review

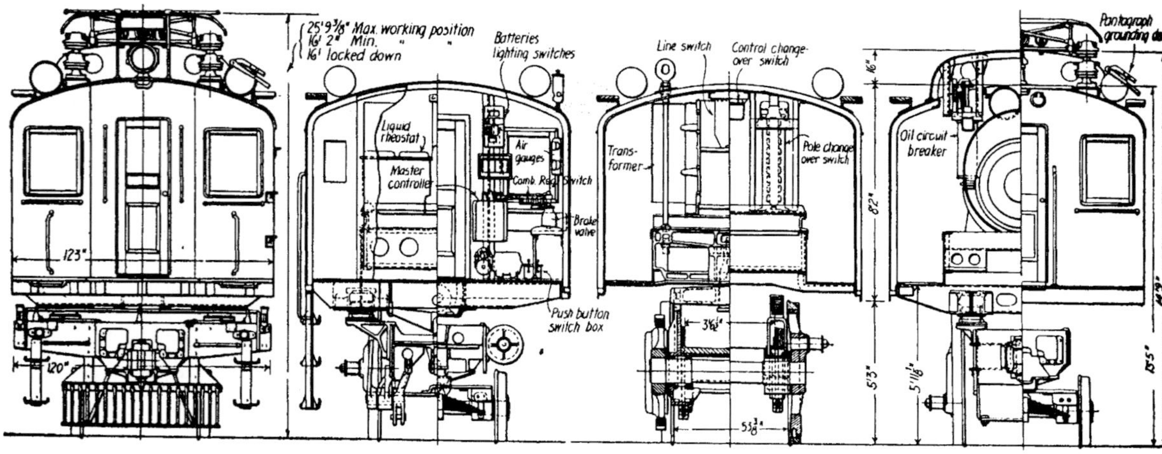

These end elevations and sectional views of the N&W Class C Electric Locomotive help to show the layout of equipment inside.

Mason Y. Cooper Collection
Originally from Railway Review

nected in parallel, but utilizing the four pole secondary windings on the induction type traction motors. The locomotives were equipped with switch controls and arranged for the operation of the two units from either control cab.

The control equipment was specifically built for handling A.C. collected from the 11,000 volt trolley wire by the pantograph. This current was fed into the main transformers, equipped with an oil type circuit breaker, which reduced the power to 725 volts. A phase converter was connected to the low tension (725 volt) side of the transformer and operated continuously while the locomotive was in operation. Externally, the phase converter resembled a big rotary motor, and to its extended shaft was coupled a blower for cooling the motors and transformers by means of the duct channel in the floor. Through a clutch on this shaft, an air compressor was also operated. The phase converter was an induction motor with a short circuited or cage-wound secondary having two windings on its stator (glossary). The first circuit drove the rotor (glossary) and the second circuit furnished current out of phase with the main supply current.

The motor primary winding circuit of the phase converter was connected across the secondary circuit of the locomotive transformer. This motor circuit received current at 725 volts. (With later technological improvements this value was raised to 750 volts.) When running, the wiring arrangement of the phase converter windings resulted in a current of 90 degrees phase displacement being induced by the secondary winding onto the phase converter's primary winding. By then connecting this 90 degree displacement circuit to the middle tap of the main transformer, a 725 volt three-phase current was provided to the traction motors. It was only necessary to use a small portion of current in this displacement, as the largest portion of current came directly from the main transformers. For starting the phase converter, a single phase commutator type motor was mounted directly on the shaft of the converter.

Details of the 2-4-4-2 boxcab locomotives included two pantographs mounted on the roof of each locomotive unit. They were unique in that they could be fitted with end horns, which would automatically fold in when the pantograph encounters a low point such as tunnel trolley wire. The trolley was raised and held in contact with the overhead wire by springs. The crew could lower it with compressed air. Double insulation was provided for the pantograph by two separate sets of insulators arranged in series. If one set should happen to fail, the other set would still be effective in sustaining operation.

On each locomotive unit were four liquid rheostats, one for each traction motor. The rheostats were operated in pairs, each pair having in common a

This broadside view of Class C 2504, taken at Bluefield, WV, shows the boxcab layout of the pantograph on the Baldwin-Westinghouse engines.

storage reservoir and cooling tower, a circulating pump and a operating mechanism. The liquid rheostats provided the motor circuit a variable resistance required so that the speed of the motors may be slow in starting, and could gradually increase as the resistance was cut out of the circuit. In other words, if the 14 miles per hour speed was selected, the rheostat would vary the resistance to allow the locomotive to start from a dead stop and smoothly accelerate to that selected speed and torque.

The essential elements of each rheostat were a cast iron tank into which were mounted iron electrodes. These consisted of metal plates, a portion of which were connected to the motor. The remaining electrodes in the form of additional plates were connected to ground. A hollow plunger was located at the center of the tank which regulated the height of the liquid. The plunger was raised and lowered by an operating mechanism into the solution of anhydrous sodium carbonate. This solution was continuously re-circulated by a motor-driven centrifugal circulating pump to prevent excessive heating that resulted from interaction with the electric current. The liquid was forced up through the bottom of the tank and over the top of the hollow plunger when the plunger was raised and back through the inside of the plunger to a storage reservoir. The liquid rheostat afforded exceptionally precise motor control and was used for the first time on these N&W locomotives.

When the engineer wanted to operate the locomotive, inside the rheostat the liquid raised to submerge the lower ends of the plates, and electric current flowed between the motor-connected plates and the adjacent ground connected plates. In this manner, the circuit between the three phases of the motor was established. The higher the liquid rose over the plates, the lower the resistance became; as a result, a larger amount of current could flow. When the plates were entirely submerged with the liquid going up all the way, the resistance was reduced to the minimum. In addition to the re-circulation by a pump, other measures were taken to keep the temperatures of the circulating liquid anhydrous sodium carbonate as low as possible. A portion of the fluid was circulated through a cooling tower containing a series of trays. These were arranged with the liquid flowing over them in a thin sheet to be cooled by air from the main ventilating duct in the floor, to which the towers were connected.

A 36 in. Sirocco fan was mounted on the shaft of the phase converter to cool various appliances. It was arranged to deliver air to the duct built into the cab floor under the deck. The duct ran lengthwise through the cab with openings at various points where connections were made to the traction motors, phase converters, cooling towers and other accessories.

The first eight of the planned twelve locomotives were delivered to the railroad by June 15, 1915. They were assigned to Class "C" and given road numbers 2500-2507. The finished locomotives had a total driving wheelbase of 83 ft. 10 in. The wheelbase between trucks was 16 ft. 6 in., with a rigid wheelbase of 11 ft. The height of the locomotive from the head of the rail to the top of the cab was 14 ft. 9 in., and to the top of the pantograph was 16 ft. 6 in. The width of the cab body was 10 ft. 3 in., while the maximum overall width of the locomotive was 11 ft. 6 in. The diameter of the drivers was 62 in. with 30 in. pony truck wheels. Total weight of the locomotive was 270 tons, with 220 tons of that amount carried by the drivers.

In tests, the locomotives were found to develop tractive effort higher than the guaranteed maximum. Windings for both four and eight pole operation with

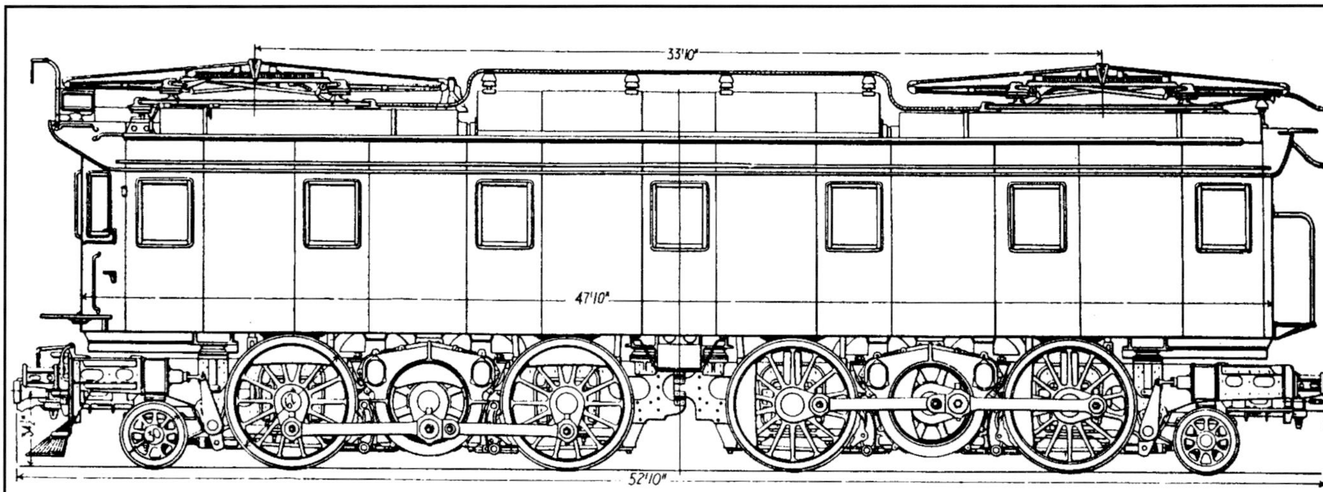

One-half of a Norfolk & Western Railway Class C Electric Locomotive.

Mason Y. Cooper Collection
Originally from Railway Review

Engine 2502 was one of the initial locomotives received to begin operations. The Class C had a normal rating of 1625 tons when operating from the coal assembly yard at Eckman over the grade to Bluefield.

provision for cascade connection enabled the constant speed induction motors to provide running speeds of 7, 14, and 28 m.p.h. Regenerative braking was also possible. In testing with N&W's Dynamometer car, on one occasion a pair obtained a maximum 180,000 lbs. tractive effort which calculated to a 40% factor of adhesion. The locomotive was tested on 1%, 1.5% and 2% ascending grades, both individually and in pairs. Tests were also conducted on the 0.4% descending grade. Drawbar horsepower during these trials ranged between 5,000 and 6,700, and varied with the speed. The railroad had in its possession a formidable locomotive. ■

LARGEST ELECTRIC FREIGHT MOTOR IN WORLD, N. & W. RY., BLUEFIELD, W. VA.

This postcard, issued shortly after the arrival of the electric locomotives, proudly announces they are the largest electric freight motors in the world. ROGER NUTTING COLLECTION

4 Electrification!

The railroad moved quickly to set up for electric train operations. Construction work began in 1913 on the powerhouse at Bluestone Junction and simultaneously on catenary erection. Work on the powerhouse, which had an initially installed capacity of 24,000 k.w., was completed in the summer of 1914. Catenary construction was carried out by a specially organized railroad force. Initial energization took place on December 17, 1914. The section from Ennis to Ruth, including Elkhorn tunnel, was energized on January 17, 1915; Bluestone Junction to Ruth on January 19th; Flat Top Yard to Falls Mills on March 15th; Bluefield yard on April 29th; and the main line from Eckman to Landgraff on June 11, 1915.

Shortly afterward, motive power road tests began on the locomotives, eight of which had been received. Assigned to

Electric power lays over at the assembly yard in Eckman, WV.

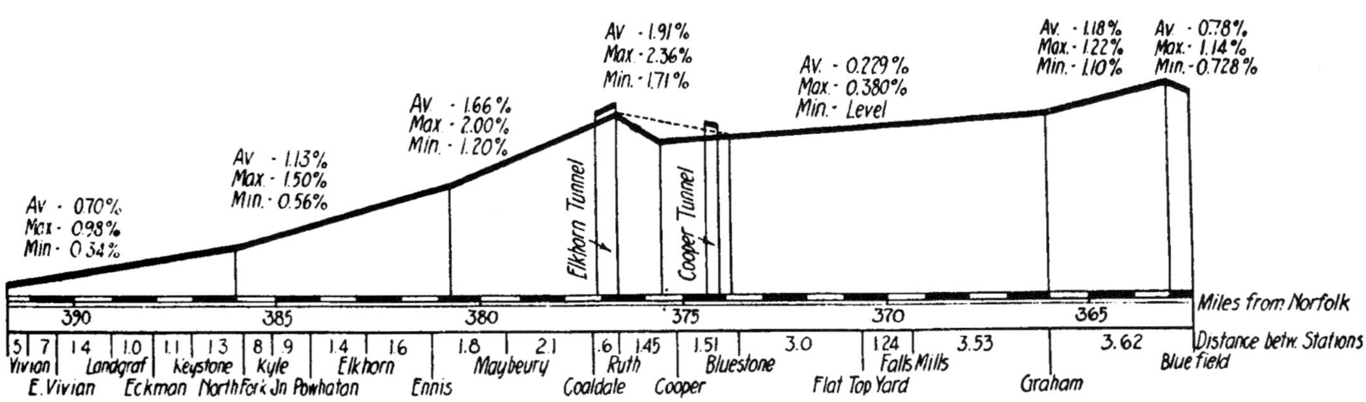

This profile of the electrified lines shows the grades against which N&W moved coal to market.

Mason Y. Cooper Collection

Class C, each unit consisted of a pair and was given road numbers 2500-2507. By June 30, 1915 the catenary had practically been completed between Bluefield and Vivian, a distance of 29.35 miles. The power plant was operational and 8 locomotives were in service. The railroad had expended to date $3,128,224.04 on electrification and looked for a return on investment.

Mr. Q.W. Hershey described early operations in an industry report dated June 4, 1915, finding many similarities between electric and the steam engines then engaged in coal service. Normal operations with steam mallet locomotives at this time, where long and heavy trains were handled, used one or more 2-6-6-2s on the head-end and one or two more pushing on the rear. Start of a train movement was controlled with little difficulty by whistle signals, allowing the pusher locomotives to take up slack and stand against the load. The head end would then begin to pull and the train moved out. The new electrified section constituted practically a separate gathering division existing within the Pocahontas region, comprising about 85 track miles in the 30 mile stretch. The section was double tracked throughout except in the Elkhorn tunnel. In addition, there were numerous passing sidings and branches into the mines. The heavy coal business originated west of the summit and for the most part was hauled east over 1.5 and 2 percent grades.

Coal traffic on the Elkhorn grade consisted almost wholly of long heavy tonnage trains which continued to increase in quantity each year. N&W's practice was to continuously upgrade all facets of its operation to obtain maximum benefit from any capital improvements. In the 1915 Annual Report, the railroad stated

Class C 2508 works a coal train up the Elkhorn grade at Maybeury, WV. A second set of electric engines is pushing on the rear.

it had added 537 all-steel hoppers of 57.5 ton capacity and two (experimental) 90-ton all-steel flat bottom coal gondolas to deal with an increase in business.

Coalfield train operation had remained a difficult ordeal with steam motive power. Once a train was underway, the crew had to deal with the associated traffic congestion and noise resulting from steam operation. There was difficulty sometimes in securing simultaneous action in the starting and braking for signals, and stops to take on coal and water. The design of the new electric locomotives took into account experiences of steam operation in that the locomotives would be able to take up slack to their full tractive effort. They could stand stationary at maximum draw bar pull for a period of up to 5 minutes against the load until the locomotives on the other end would come into play, and still encounter no damage. Once underway, the electric engines took advantage of their increased speed to stay out of the way of other traffic.

Initial impressions from the use of three-phase current in the A.C induction motors was found to be rugged and trouble-free. The locomotives had a rated one hour capacity of 3,000 HP. Constructed without commutators or commutating devices, the A.C motors were not subject to sparking, burning, pitting, or brush troubles under heavy load. The problems experienced in replacing steam with earlier D.C. electric operations had been eliminated.

Overall, conditions in the N&W coalfield were found favorable for electric haulage. This consisted primarily of collections from the mines, sidings and yards, then hauling this heavy fixed tonnage up and over the grades into the eastbound classification yard at Bluefield. Coal traffic was dense and moved at fairly uniform intervals. Coalfield operations could be run practically as a separate engine division as there was no intention to conduct through merchandise or passenger service by electricity. This service was to be confined to traffic designated as "tonnage trains" and was designed with a capacity of about twenty trains a day,

Electric engine 2506 exits the east end of Elkhorn tunnel at Ruth, WV. Steam motive power operated at reduced capacity in winter weather while electric operations were largely unaffected.

or 650,000 tons.

Norfolk & Western's purpose in electrifying this part of the railroad was (1) to increase the capacity of the line by materially reducing the time required to handle trains, and (2) to provide a more economical service over the Elkhorn grade. To this end, the heavy coal trains were handled by electric locomotives at a running speed up the grade of 14 m.p.h., as compared with about 7 m.p.h. under steam operation. A further time savings was effected by eliminating the delays of normal steam operation: occupying track space at the several coal and water stations on the grade.

The outcome of this increased speed was especially marked at the single track Elkhorn tunnel. Due to ventilation requirements, it had been necessary with steam operations to reduce speed upgrade in the tunnel to about 6 m.p.h. With steam locomotives, it required seven minutes to clear the block at the tunnel, whereas under electric operation this movement was made in about three minutes. Coal operations moved in about 12 trains on the average day during the summer of 1915, with additional helper and pusher service supplied for through freight and passenger trains.

A report dated February 24, 1916 by Mr. Q.W. Hershey reported on the experiences of the railroad during the first winter of operation. Under electric operation, weather conditions did not necessitate reductions in train tonnage, as was the case with the declining thermal efficiency of steam. The normal trainload of 3,250 tons continued to be handled with two electric locomotives running at 14 m.p.h., an improvement to operations. Coal used at the Bluestone Junction powerhouse was running about 12,000 tons a month. When taken into account with the coal used by mallets performing the same amount of work, this resulted in a monthly savings of $30,000. In addition, light priority time freights and passenger trains were also assisted by electric helpers. These now operated at the higher 28 m.p.h. where 10 and 20 m.p.h. were the practice under all-steam power. Passenger train schedules were adjusted to reflect the twenty minute savings in time.

Train dispatching under electric operations was now found to be simplified as schedules could more definitely be determined. This was because electrically powered trains moved at a fixed speed. The uncertainty with steam power due to coal and water stops was eliminated. No trouble was experienced in training the former steam locomotive crews to operate the new electric equipment. The operating routine remained the same and the simplicity of the electric controls made the changeover easy.

On June 30, 1916 the railroad released figures which showed just how successful the new electric operation had been. Freight revenues had increased over 35%, an increase of $13,008,590.00 over the previous year. Net income increased an astounding 98% to a total $20,624,058.57. Four additional Class C electric locomotives were received and assigned road numbers 2508-2511. This brought the fleet to a full compliment of 12 locomotive sets. In order to use the new locomotives to the line's newly expanded capacity, the railroad added 371 of the all-steel 90-ton flat-bottomed battleship gondolas and 300 of the 57.5 ton all-steel hoppers to the rolling stock roster.

All twelve electric locomotives had been delivered during 1916. Of these, one unit was set aside for repairs from an early wreck. Regular operation was restored with nine of the motor sets in use, and the balance rotated through for maintenance. The Class C engines were not without problems, but this was not unexpected. Due to the pioneering aspect of the installation, mistakes were made in the design by the manufacturer and in specifications from the railroad. The buffeting and shock from slack of heavy loads took its toll on the engines. The bulk of the problems could be traced to an under design of the components. The most serious problem was with frame breaks, which could not be fixed without major overhaul. By 1921 about half the fleet required new frames, with the work done at Bluestone shop. The cost of the work was contested but split between builder and railroad.

Another problem, known as "chattering wheel slip," was caused by slippage between the gear on the traction motor armature shaft and the pinion on the jack shaft. The cause was traced to misalignment of components resulting from the aforementioned frame damage. Once this condition began, it usually became more frequent and severe until frame replacement took place, due to the continued breakage of teeth on the pinion. The only remedy was to pull the locomotive unit from service and perform the frame change out, then replace the damaged pinions. Otherwise the traction motor mounts continued to work loose in the everyday course of getting over the road.

Both frame and slack damage caused breakage of the cast casings for both the oil-type circuit breakers located in the car body and traction motors setting over the jack shaft. This allowed oil from the circuit breakers to get into the armature windings. Oil also dripped down from the gears onto the track to such an extent it resulted in wheel slippage. It became necessary to install catch pans on the engines to keep oil off the rail. Still other problems developed with the crank pins and side rods which were simply too light for duties. Replacement and eventually redesign were necessary, the cost again shared between the railroad and the manufacturer. Still, many locomotives worked to the end of their lives with clanking siderods from heavy wear.

One problem which soon became apparent was the excessive amount of heat generated by the units when under load. During the summer months, this could become quite uncomfortable. Often crews would leave the windows and cab doors open to improve air flow. Excessive heat also plagued the catenary. Soon after operation commenced the size 00 auxiliary messenger wire which supported the trolley wire, and the size 000 trolley wire itself were overheated on heavy grades on account of the enormous current demands. To remedy this, a second auxiliary messenger was added at certain points.

Aside from these matters, a maximum of eleven engines were available for service. In a normal working day, an average 80 miles per day was made in a combination two round trips bringing a total of 90 loads into Bluefield per train.

With the mainline electrification between Bluefield and Eckman yard now in operation, the railroad began a program to extend the catenary. First to come under the wire were the branches east of the mountain feeding coal loads into Flat Top yard. Extension of the electric lines from Bluestone Junction was begun over the 1.52 mile branch into Pocahontas, and up the Bluestone branch. The wire was extended up the branch to within one mile of the Simmons Creek spur. This section was activated September 1, 1916. The catenary into Pocahontas, VA was activated on November 6, 1916. The delay was due to the necessity of placing the yard located here under wire. Electric locomotives took over movement of coal off the branch.

With electrification proving a success, another extension known as the Wilcoe-Farm Extension was approved, bringing an additional 16 route miles and 51 single-track miles under the wire. The move was intended to bring about an additional reduction in operating expenses by placing a larger portion of the coalfield under electric operations. There were 248 coal mines now located on the road, with 245 of these in production. The extension was broken down as follows; 11 miles of double track mainline from West Vivian to Farm, and 5 miles of branch line from Tug Fork Junction up to the coal assembly yard at Wilcoe.

The catenary was extended up the Bluestone branch past Bramwell, WV to bring additional mining activity into the realm of electric operation.

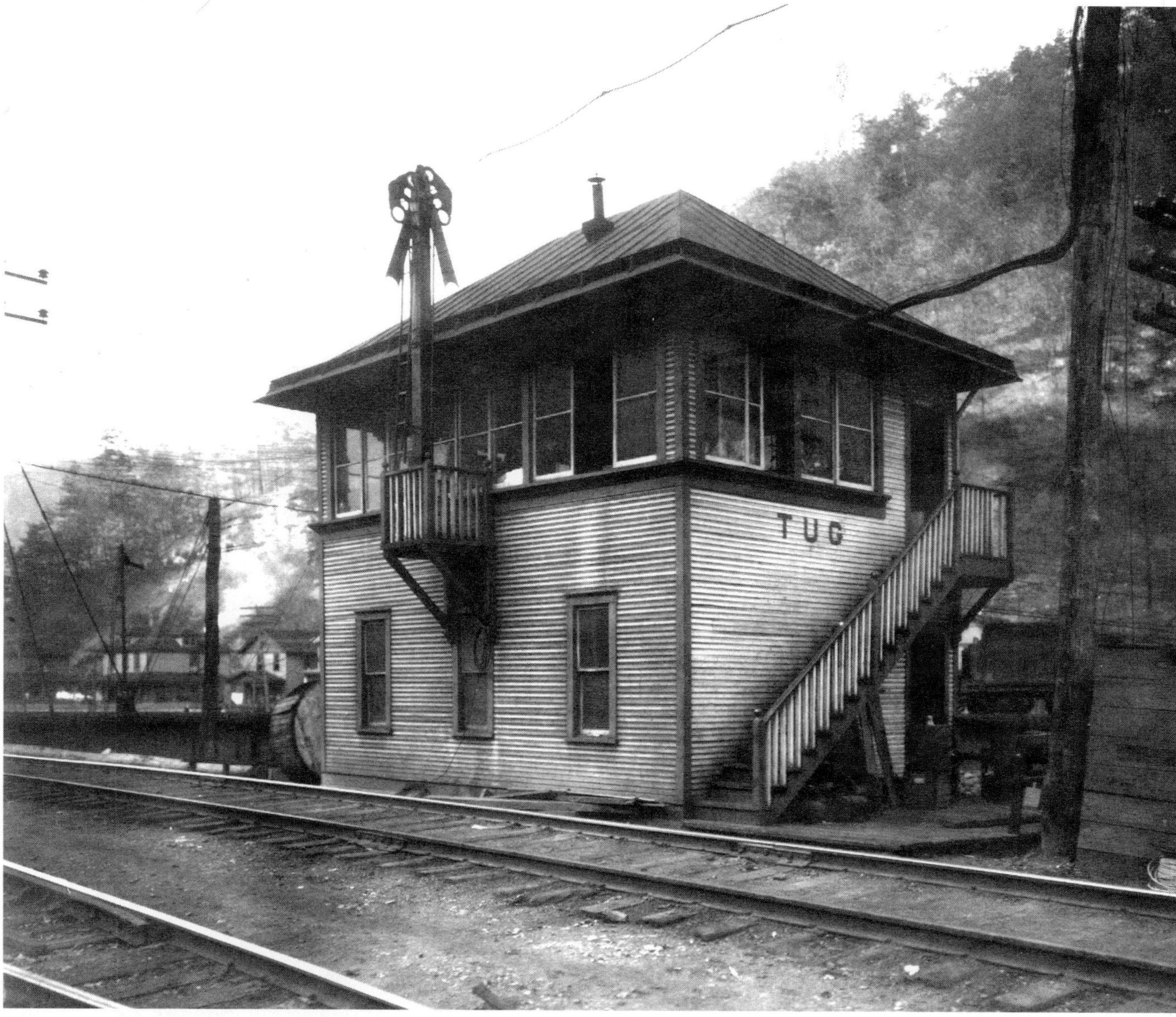

Work to extend the catenary past Tug Tower at Welch and westward to Farm was suspended by a scarcity of materials during WWI. The extension was completed shortly after the conflict.

A significant portion of the coal mined in this area moved eastward. Demands placed on critical materials by the war in Europe were expected to prevent completion of the project before January 1, 1918.

Indeed, the war was beginning to effect rail operations in another way. Added traffic was beginning to impact the railroad as a whole with dire results. In a report dated December 31, 1916, a 10.8% increase in operating revenues translated into a 13.9% decrease in net income, which fell to $18,946,137.04. This was due to higher costs of operation, at least in part due to the increased demand for coal. Additionally, an inefficient utilization of equipment resulted as large numbers of loaded cars sat in port areas along the east coast waiting to be unloaded. The operating ratio increased from 56.36% to 62.45%.

Coal loading continued to increase throughout 1916 to meet war demand. Between 1906 and 1916, the eastbound coal net-ton haul had increased from 279,933,339 ton miles to 592,597,018 ton miles, an increase of 110%. Coal prices increased sharply with the railroad obligated to purchase lands in Mingo County, WV and Pike County, KY in order to guarantee a supply for it's own locomotives. Fifty all-steel 90-ton flat bottom battleship gondolas and 438 all-steel 57.5-ton hoppers were added to the rolling stock fleet to accommodate the traffic. Material shortages continued to worsen, with an experimental 57.5-ton wood and steel composite hopper, and a 100-ton all-steel battleship gondola received. Both experimental and all regular gondolas and hoppers added during the year were constructed in Roanoke Shops for use in the Elkhorn field.

The electrification on the Bluestone branch was extended up to the Simmons Creek spur to relieve the obligation of using steam in this area. Mining activity had pushed this line another two miles on separate spurs with 4 active mines and 3 coke oven operations. This section was activated on November 15, 1917. Mainline electric construction continued although progress was delayed by scarcity of materials, especially copper. Completion of this portion of the work was delayed until late 1918.

The national state of railroading continued to worsen as traffic problems and car shortages plagued the carriers. President Woodrow Wilson seized control of the railroads effective December 27, 1917. The Federal Control Act, passed March 21, 1918, established the United States Railway Administration (USRA) to assume operation of the national rail system. The USRA provided the N&W an aggregate annual compensation of $20,634,142.03, based on the average operating income for the railroad for the three years ending June 30, 1917. Roanoke shops constructed 1043 57.5-ton wood and steel composite hoppers during the year for coal service.

Further work on the Wilcoe-Farm Extension was stopped by the USRA administration. Reasons cited were delays in obtaining new electric locomotives and scarce labor conditions brought about by the war. It was estimated that an additional three months of work remained in order to complete the present extension from West Vivian to Farm, and from Tug Fork Junction up the branch to Wilcoe Yard. Under USRA direction, the electric locomotive fleet was re-directed to move regular freight as well as coal trains. Under this operating scheme, the number of trains moved by electric power increased by 44%. By 1919 operating revenues had fallen by $5,078,435.22 through a combination of inefficient operations and a 18.93% decrease in freight tonnage.

With the world war coming to a conclusion, President Woodrow Wilson announced the end of federal control of the railroads on December 24, 1919. The changeover was scheduled to occur on March 1, 1920. Throughout the electric operating zone an examination by N&W of the right of way found excessive wear to rails and cross ties that occurred during USRA tenure. Motive power, both electric and steam, showed wear from hard use. Net income for the USRA period as reported on December 31, 1920 was sharply lower ,driven by federally mandated 31.5% payroll and 80.8% fuel increases.

Coal traffic continued to expand. Under USRA direction, Roanoke shops had turned out 88 additional wood and steel composite coal hoppers. These were put to work immediately in the Pocahontas field to deal with increased industrial demand. Following return to private control in 1920, the railroad authorized additional expenditure for rolling stock, including 444 all-steel flat-bottomed battleship gondolas.

An examination of electric train service found operating costs had gotten out of hand. The total cost was discovered to be $237.31/100 locomotive miles for electric operation against $78.50/100 locomotive miles for steam operation. These figures included all charges for fuel, lubricants, stores and costs of repair. The reason was determined to be a misapplication of motive power by USRA.

N&W had originally intended to use the electric engines for heavy coal train movement. In this dedicated service, the electric locomotives replaced several steam engines in high capacity movements and made more trips per day. In these circumstances electric operation was an economical alternative to steam. Under USRA operation, electric motive power was re-directed away from dedicated coal train service to work on regular freight trains. This "next train - any power" philosophy had an unforeseen effect on costs. Tonnage carried by these movements was often far below the capacity of the electric locomotives, while costs remained fixed. In 1921 the railroad reverted to its prior operating philosophy and charges to the electric locomotives fell by over $180,000.

In addition to the inefficient use of motive power, the huge volume of war-time traffic again resulted in congestion at Elkhorn tunnel. Traffic simply could not clear the section in sufficient time to allow for regular train movement. A study of the problem revealed that eastbound trains were again taking an inordinate amount of time to clear the tunnel block. The problem was due to the single track bore and the time consuming eastbound descent at 12 miles per hour down the steep grade to the Bluestone River at Cooper. The resulting bottleneck slowed the movement of all traffic past this point.

In November of 1921, the railroad completed a study recommending a series of improvements intended to relieve the congestion. A second tunnel was proposed between Ruth and Coaldale, located 200 feet to the north at a slightly higher elevation. This new tunnel was to be utilized by westbound traffic, and tied into the existing right of way at the crossovers located 5,000 feet to the west of Coaldale. The plan also called for replacement of the existing eastbound mainline between Ruth and Bluestone Junction by construction of a second line parallel to westbound mainline. This westbound mainline had been constructed at a later date and benefitted from a more favorable alignment.

This new construction would first necessitate the enlargement of Cooper Tunnel to accommodate a second track. This part of the project was undertaken first and proceeded during the movement of traffic. These improvements would allow eastbound traffic to move

An electric train brings coal out of the Elkhorn valley in the 1920's. A second set of locomotives can be seen in the distance assisting on the rear end.

at the higher speed once clear of the tunnel and as a result faster clearing of the block. Analysis of the plan led to approval of reconstruction of the eastbound track but rejection of a second Elkhorn Tunnel. It was determined the general speed-up of traffic resulting from improved movement east of the tunnel would eliminate the need for a second bore under Flat Top mountain. In addition, the economy was beginning to slow, and the effect on future traffic was at this time unknown.

Despite the business recession and labor difficulties, both mining and coal loading continued to increase. The composite wood and steel hopper cars received during USRA control showed excessive wear, prompting the railroad to order more all-steel hoppers. A strike of shop forces called for July 1, 1922 led N&W to place orders for these cars with commercial builders in order to insure the ability to meet customer demand. A total of 5,503 all-steel 70-ton hoppers were delivered to the railroad. Also, 1,393 of the composite hoppers were put through a rebuild program, emerging as all-steel cars. The extension of mainline electrification was completed during 1922. Eastbound coal train operations out of the assembly yard at Wilcoe now moved with electric motive power.

There were now 325 coal operations on the N&W system with 222 in actual operation. The center of mining activity gradually moved west from the Elkhorn region. The Dry Fork branch and locations along the mainline west towards Williamson came into a high state of production. The railroad added several small short lines and branches to keep pace with the growth. During 1920 N&W had leased the Tug River & Kentucky Railroad, a 4.45-mile line which the railroad operated as its Cedar, McCarr and Delorme branches. The Williamson & Pond Creek Railroad was leased to become the Pond Creek branch. In a joint agreement with United Thacker Coal Company, the railroad undertook the construction of 18 miles of line. This ran from Lenore, WV in a northeast direction up Pigeon Creek to Thacker Coal properties in Mingo County. On October 12, 1923, the company acquired all the capital stock of the Big Sandy & Cumberland Railroad, which owned and operated 33.08 miles of lumber road from Devon to Grundy, VA. The area contained vast reserves of coal. N&W also acquired the stock of the Knox Creek Railroad, which was leased to the Big Sandy & Cumberland. This company owned the right of way, which extended westward into Kentucky.

Further expansion of coalfield electrification was authorized in 1923 with the announcement of the Iaeger Extension. Scheduled completion for the project was to occur by the summer of 1924. Construction work began at Farm at the end of the wire, and worked west a distance of 22 miles to beyond Iaeger, to a point on the center of the siding known as the Hull Middle Track on the N&W main line. The project included a branch running southeast from DY tower, on the south end of Iaeger, for 1 mile to the south end of Auville Yard. Auville Yard was the assembly point for coal traffic off the Dry Fork branch. When the electric extension was placed into operation, the railroad would have 60 route miles under the wire, or about 200 track miles including all double tracks, yards and passing sidings. It was expected that within the following few years, the entire coal region from Bluefield to Williamson would be equipped for electric operation.

At this time, practically all coal mined on the N&W going into Bluefield was handled from the assembly yards by electric locomotives. The exception was traffic routed via the Clinch Valley line. The electric operation had gained a worldwide reputation for heavy tonnage and efficiency. The Electric Department at the Bluestone powerhouse was visited by railroad representatives on tour from Japan, France and Chile in May of 1923. N&W President N.D. Maher opined in *Electric Railway Journal* that "N&W might also electrify as far as Roanoke with good assurance of an excellent return of investment."

Concurrent with the electric extension was the need for additional motive power. In early 1922, Norfolk & Western engaged Gibbs & Hill to determine what course of action should be taken with respect to the problems which had occurred in the initial electric locomotives. Experiences gained through

Electrification was extended past Davy in 1924. The supporting structure of the catenary was modified with a lightweight truss arrangement replacing the center steel H-beam and the round support columns.

road service had revealed frame weakness, which in turn led to damage in auxiliary locomotive systems. Gibbs & Hill determined the best solution would be to purchase new engines taking benefit of advances which had taken place in the industry since the construction of the original locomotives. Special attention was paid to addressing the frame problems brought about in N&W's heavy tonnage mountain operation.

Work was soon underway. On September 9, 1922, Gibbs & Hill announced that four new electric locomotives were being ordered for the Norfolk & Western Railway. The heart of the new engines would be 16 specially prepared vanadium steel side frames. These high-test frames were supplied by Union Steel Casting in conjunction with the Vanadium Corporation of America, who had developed a new high-test steel for locomotive applications. Minimum tensile strength of this new steel was 10,000 pounds greater than for standard vanadium steel. Each of these N&W frames were 41 ft. 2 in. long with the main section 6 in. wide, weighing 23,700 lbs. Each engine unit required one frame under each side. Since two units made up one locomotive, 4 frames were needed for each of the 4 locomotives ordered. Steel for the frames was prepared in either open hearth or electric furnaces with low tolerances for chemical purity. This steel was then used to prepare the castings. It was allowed to cool, then heated again to a specified temperature to refine the grain. Special provisions were made to anneal, which made the frames less brittle, and re-anneal the castings if necessary. Only minor imperfections could be addressed by welding, and then only after the approval of an inspector. Otherwise, the frame was scrapped and a new casting prepared.

The new engines, initially classified C-2 like the original Class C motors, consisted of two cabs permanently coupled to form a single locomotive, and were one-third larger in capacity than the original electric locomotives. Construction was to be carried out by Westinghouse and American Locomotive Companies. The equipment in each cab was identical so that when coupled back to back the locomotives could be operated from either end like the original Class C power. The design of the cab and running gear differed as a result of a combined study by railroad and Gibbs & Hill. The units were constructed in such a manner to meet the requirements of severe mountain service.

The cab structure was fastened to and carried by the side frames. This contrasted with Class C engines, in which the cab was supported by springs and sliding bearings. In the new locomotives the vanadium steel side frames were connected to cross ties which were used to support the heavier pieces of electrical appliances that were mounted in the cab. The four pair of drivers in one cab were mounted in a single truck, whereas in the prior motors, two trucks of two pairs of drivers each were themselves connected by a mallet hinge. The spring suspension arrangement of the 62" drivers were similar to the Mikado steam locomotive type. It embodied side equalization between the lead-

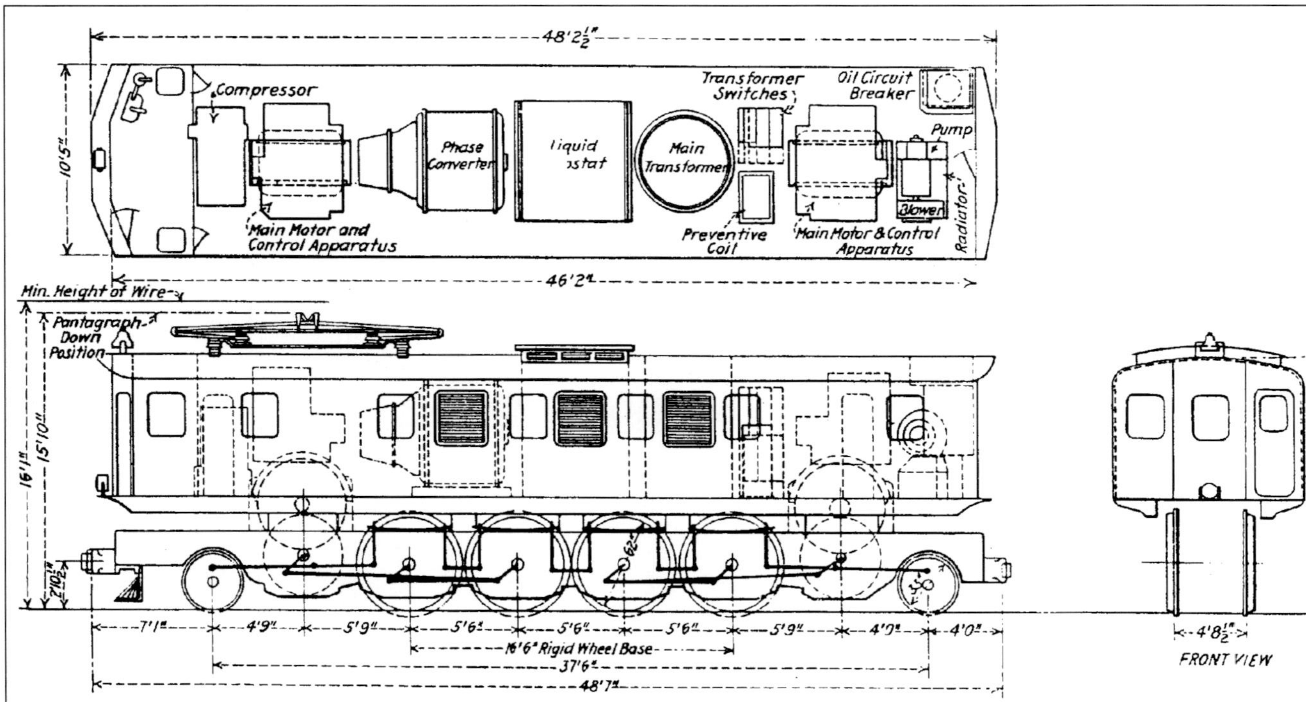

This side profile of the new electric locomotive shows the interior location of some of the equipment.

MASON Y. COOPER COLLECTION
ORIGINALLY FROM RAILWAY AGE

The first member of Class C-2, the 2512, pauses at Bluestone, WV. The 2-8-2+2-8-2 wheel arrangement differed from the earlier C-1 engines.

This head-on end view of C-2 2514 shows the revised placement of the headlight and the dual purpose marker classification lights that were, over time, applied to all electric locomotives.

ing truck and the adjoining two axles, as well as cross equalization between the other two driving axles and the trailing truck.

One large motor, rather than the two smaller motors on the earlier units, drove each jackshaft. Maintenance difficulties of the original engines resulted in changes to the jackshaft design on the new engines. The jackshaft and its bearings were now more easily removed. The jackshaft was carried in a heavy bronze bearing, which was split vertically and rested on a steel collar casting. These were quickly changed out resulting in less down time for the locomotive. The bearing and collar casting sat in the side frame between fitted pedestal ways. The collar casting itself was provided with a tie bar arranged so the collar became almost an integral part of the side frame like the old engines, only now the whole apparatus could be more easily repaired.

A redesign of the new locomotives was made to deal with the pinion problems. A flexible gear mounted on each end of the jackshafts now transmitted the motor torque by means of a side rod to the wheels. The springs were inserted without initial compression, and were designed not to become fully compressed until a torque was exerted equivalent to 110 percent of locomotive adhesion.

The design of the main motor closely followed specifications laid down by Westinghouse for an experimental split-phase locomotive for the Pennsylvania Railroad. Westinghouse had put considerable work into a planned Pennsylvania electrification of the Altoona grade, hoping to sell a larger engine using a total of eight motors instead of N&W's design which used four of these motors per unit. Additional refinements had taken place in the basic motor design itself, since the first N&W split-phase locomotives were delivered in 1915. More powerful units were now possible. Specifically, there were improvements in ceramic insulation, which allowed for thinner motor walls. This meant additional space was now available to put more windings on the rotors, which in return made more powerful motors. Structurally, the induction motor was similar to the prior N&W design, in that it had a wound secondary arranged so that it could be connected for either four or eight poles, corresponding to the two speed operation. Collector rings were now provided to more surely connect the secondary windings to the liquid rheostats which provided the starting resistance. These connectors were mounted on the ends of the motor shaft to provide quick accessibility and ease of inspection.

The nominal rating of the traction motors was 1,000 H.P. each, or 4,000 H.P. per hour. The four motors on one complete locomotive were capable of developing 90,000 lbs. tractive effort continuously at 14 miles per hour, and 52,000 lbs. continuously at 28 miles per hour. The one hour rating of tractive effort increased to 108,000 lbs. at 14 miles per hour, and 63,000 lbs. at 28 miles per hour.

The stator in the synchronous phase converter was wound like a two- phase induction motor. Indeed, the apparatus resembled a big round motor. The phase converter was wound for direct current excitation. In order to prevent overload, locomotive power was controlled so that at full load it was operating at 95 percent capacity. The phase converter was started by means of a small single-phase commutator type motor mounted directly on its shaft. It was arranged so that when the starting duties had been completed, it reconnected as a separately excited D.C. generator to furnish current for the phase converter motor. Separate excitation for the D.C. generator was obtained from the control generator and storage battery.

The phase converter stator windings were arranged so that one of them acted as a drive winding for the motor. The other winding had a voltage generated in it which had approximately a 90-degree phase relation with the transformer voltage. By combining this voltage with the transformer secondary voltage in accordance with the usual two-phase or three-phase connection, balancing three-phase voltage was produced at the traction motor terminals. By means of a balancing transformer preventive coil, normal motor voltage was maintained even with a 15 per cent drop in the voltage received at the trolley wire during acceleration. The coil also worked with a drop in half of voltage at the trolley wire during normal running.

Auxiliary motors on the locomotives were designed for easy replacement and interchangeability. Cooling of the oil-insulated transformer was accomplished by means of pumping the oil through a forced ventilated radiator. An auxiliary motor supplied the energy to drive both the oil pump and the blower for the ventilating air. A motor-driven blower set for each motor accomplished ventilation for the traction motors. The oil pump, ventilating pump and the pump circulating electrolyte in the liquid rheostat were identical. The only difference in the last motor was that it had a shaft extension. Otherwise all the auxiliary motors on a locomotive were interchangeable.

Following prior practice established with the earlier Class C engines, the generator that furnished the energy for control, battery charging, and D.C. excitation of the phase converter was chain driven off the shaft of the phase converter. It had enough capacity so that in the event of a control generator failure in the adjacent cab unit, one generator could supply both exciters.

Westinghouse's liquid rheostat design was materially improved over the Class C engines and any other application then in use in the country. The rheostats for both motors of one cab were contained in a common tank and storage supply. This insured the same electrolyte strength in each rheostat. By means of a motor-operated common valve controlled from the master controller, the electrolyte in both rheostats was raised uniformly and at the same rate. By control arrangement, the electrolyte valves in the two cabs could be regulated so that an equal division of load on all four motors was guaranteed.

One of the outstanding features of the original Class C locomotives was wheel slip control arrangement. In the event one of the trucks of the locomotive lost adhesion, full torque could be maintained on the other three two-axle trucks holding the rail, while the torque of the fourth truck was reduced until slipping stopped. This occurred when the torque of this truck was raised to a value equal to the others and acceleration continued. This was accomplished by means of four push buttons located close to the master controller by which the engineer could lower the torque on any motor of the locomotive. By hitting the button, the torque of this motor was raised automatically to that of the others.

This mechanism was retained in the C-2 engines, as was the pneumatic cam switches which operated the traction motor circuit switch for the 14 mile per hour and 28 mile per hour connections. The control was arranged so that during transition from one speed to another, the torque of only one-half the locomotive was momentarily lost, changing over one set of traction motors at a time.

A motor-driven compressor was provided on each cab (two per engine). The compressor had a displacement of 150 cubic feet of air per minute against a pressure of 130 pounds per square inch. This motor-driven compressor set was designed for continuous operation and had sufficient capacity to charge the long coal trains operated by N&W. An air reservoir with 125,000 cubic inch capacity was provided on each cab as a safeguard.

Classified C-2, the engines were jointly constructed by American Locomotive Company and N&W's traditional electric supplier, Westinghouse. The locomotives were due for delivery about January 1, 1924, by which time it was expected the electric extension would be completed. Being more powerful, it was anticipated their principal assignment would probably be pusher duty. The new motors were assigned road numbers 2512 through 2515, falling into sequence behind the earlier locomotives which were reclassified to become Class C-1 engines.

R. J. Cousins, General Foremen of the Bluestone Electrification, was sent to observe construction of the new engines in order to become familiar with the power. He filed a report in June 1923 on locomotive construction. Alco was building the trucks, cabs and running gear while Westinghouse was building the electrical apparatus. The locomotive unit shells were to be shipped to Westinghouse in Pittsburgh, where final assembly would occur.

The report from R.J. Cousins outlined several specific points which differentiated the engines from the C-1 Class. The main transformer was oil insulated, a departure from accepted practice for heavy electric traction in the United States. As with the C-1 locomotives, there were to be two liquid rheostats per cab with one for each traction motor. These were to be connected to a common electrolyte tank. The new type of phase converter used on these engines was to be a wound rotor, synchronous type. It was designed so that the rotor would be excited from the starting motor after the converter was brought up to speed. By using this new arrangement, a locomotive loading factor of 95% efficiency under full load was achieved.

Even the box cab was of different design, now serving as merely a cover for the equipment rather than offering any structural support. All the electrical apparatus was mounted directly on the truck frames, instead of on separate supports as in the Class C-1 cab. The roof of the C-2 cab was built in sections which could be removed to facilitate repair of large interior pieces of equipment.

Both R. J. Cousins and E. L. Peters, Assistant Foreman of the electric locomotive shop, went to Pittsburgh to inspect progress on the engines in May 1923. The company then began rotating a number of shop personnel to Westinghouse in order to familiarize themselves with the new engines. N&W Machinist J. R. McLemore inspected the 4 locomotives undergoing final assembly during June of 1923. Work on the locomotives continued at a slow pace over the summer and fall. R.J. Cousins was dispatched to Alco at Schenectady, N.Y. in October, 1923 to inspect construction which had fallen behind schedule at that facility.

Delivery of the new power began in 1924, with setup for operation occurring at Bluestone Junction. The C-2s had a different look; they were shorter but heavier and more powerful than the C-1s. Each locomotive weighed 416 tons, of which 300 tons were placed on the drivers. The differing arrangement of the jack shafts gave the locomotives a compact 2-8-2 + 2-8-2 wheel arrangement, quite similar to the road's Class Y 2-8-8-2 mallets.

Trials of the new engines began during the summer months using the dynamometer car under the direction of Engineer of Tests, H. W. Coddington. Assisting were Mr. Thompson and Mr. Jenelle. Although each locomotive was shorter, it was more powerful. At 14 miles per hour, the C-2 developed 90,000 lbs. tractive effort, with a one-hour time rating of 108,000 lbs. and a short time rating of 162,000 lbs. A continuous rating of 3,330 H.P. and a one-hour rating of 4,000 H.P. was realized at 14 m.p.h. At 28 m.p.h tractive effort fell off to 52,500 lbs. with a one-hour rating of 63,000 lbs.; but horsepower rose to 3,880 continuous, and a one-hour rating of 4,660. Other individuals from N&W's electric department rode the test car including General Foreman R. J. Cousins, C. F. Taylor, E. L. Peters, and J. R. McLemore. Also along were Mr. Branson and Mr. Lawrence of Westinghouse and Mr. Goldstone of Gibbs & Hill.

During the runs, the locomotives were tested to maximum ability, and some teething problems arose. Failures resulted in various parts broken under load. These were removed and turned over to Mr. B. Cook, Assistant Engineer of Tests and K. L. Trent of the Engineering Department for further refinement. The parts were redesigned and remanufactured for retrofit on the entire class.

Mr. H. W. Reynolds was appointed Foreman of the

LC-2 2514 lays over at Bluefield. In road tests, the engines were found to be an improvement over the Class C locomotives.

Electric Locomotive Department, reporting for duty on May 6, 1924. Mr. Reynolds had joined the railroad in 1905 as a mechanical engineer, distinguishing himself in a number of positions. He designed steam engines and turbines and was later given responsibility for the design of all steam locomotives and rolling stock. In 1919 he was appointed mechanical inspector of steam and electrical equipment. As part of his duties, he assisted in the design of the C-2 electric locomotives. At the time of his appointment, he had overall responsibility over their construction.

Other electric work was underway besides the Iaeger extension. The railroad had earlier begun work to rebuild the tunnel at Cooper to accommodate double track according to plans. This project involved the relocation of a portion of the track at the time the second mainline was installed, reducing the maximum curvature from 10 degrees to 5 degrees. To accomplish this, it was necessary to relocate the catenary structure for a distance of over two miles. New foundations had to be installed which involved cutting off the poles and placing them in base castings. At some catenary structures, the cross-members had to be lengthened. In addition, signal locations had to be changed and new structural anchors erected. Steelwork on the new Bluestone River Bridge was completed first, at which point track-laying proceeded. This section between Ruth and Bluestone Junction had remained a bottleneck to operations. Passenger trains were restricted to 25 mph, and freights to 12 mph on the old line on account of curvature. Electrification of this new section required a new wire gang to be established, with Mr. M. T. "Bevo" Nunnery promoted from duties at Bluestone Junction, and placed in charge of the project. By October, 1923, work was nearly completed.

Company forces worked on the mainline catenary extension under the direction of representatives of Gibbs & Hill, operating in a specially assembled construction gang. The project reached Davy, WV, several miles west of Farm, in November, 1923, where a construction yard was established. About 250 men were hired, not including those required for work train duties. The construction foreman in charge of the project was T. B. Ogle, with R. P. Winton designated catenary engineer. The program was carried on without incident, with special precautions taken to avoid delay to revenue movements on the road.

Materials accumulated for the electrification project included nearly 90 miles of steel wire of various sizes, 223 miles of copper and bronze trolley transmission wire, huge amounts of steel for the transmission towers, and nearly 9,000 tons of concrete for the tower bases and substation pads. The overhead construction followed the general lines adopted earlier, but varied in a few detail changes. In the place of tubular poles, columns of H-beams were used. The H-beams were four different sizes: 12 in by 12 in., 10 in. by 10 in., 8 in. by 8 in., and 6 in. by 6 in., depending on local requirements. They were attached and supported on steel base plates that were bolted onto concrete foundations.

Cross-members running between the support towers were H-sections, as in the original installation. On the later Wilcoe-Farm extension, a variation in the

This postcard view looks west along the N&W main line where the old line on the left side diverges from the low grade line, located on the right. In 1924, this area was changed considerably with the relocation of the double track main via the low grade line, involving relocation of the catenary structures and the easing of curvature. The old mainline on the left then became the Bluestone branch.

Roger Nutting Collection

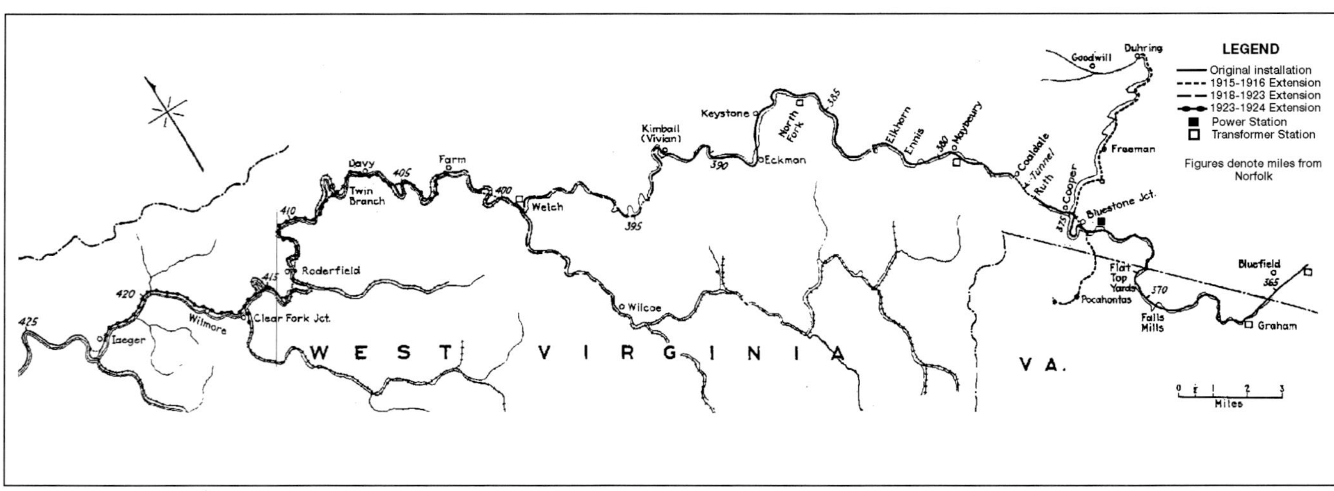

The Iaeger Extension of 1923-24 extended the electrified mainline nearly 60 miles from Bluefield. *Gary Rolih Collection*

cross-members had taken place. Due to wartime material shortages, the cross-members were made up of two angles riveted together to form a T, with spacers between the vertical sides.

Another variation was the replacement of rigid rod guys with stranded guys between the cross-members, support towers and the ground. At ground level, the guys terminated in a threaded rod and turnbuckle. Testing showed that the strand would break before pulling out of the rod and turnbuckle. The guy anchor (dead-man) was concrete over which a sleeve of iron pipe was fitted in order to offer extra protection to the rod and lower part of the strand. The guy anchors were pre-cast in company shops, then shipped to the work area.

In another variation, suspension insulators were

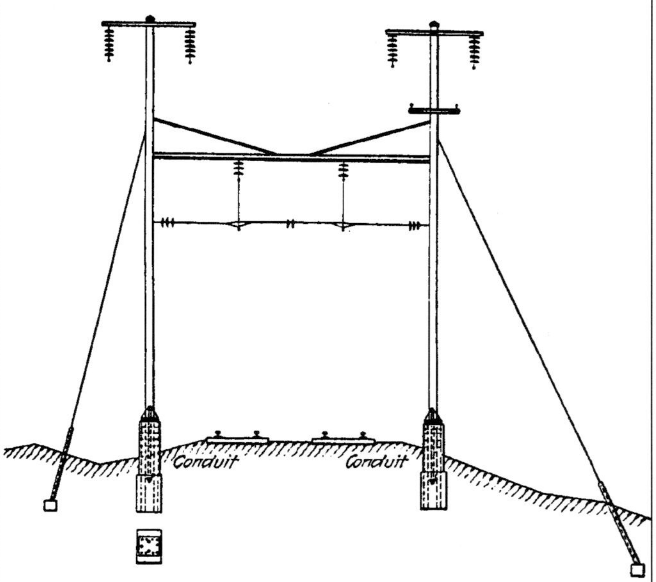

Essentials of the Farm-Iaeger Section Overhead Construction on Tangent Track Allows for Raising the Contact Line

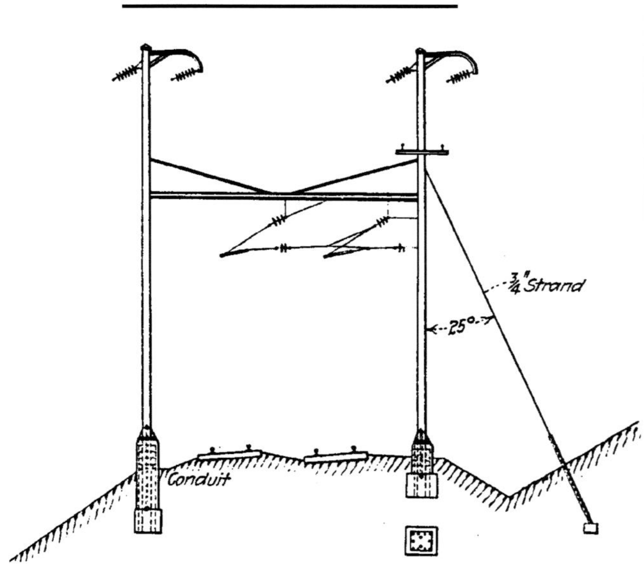

On the Curves, Which Are Numerous, Special Construction of the Overhead Is Necessary

Catenary structure differed in appearance from the original installation, taking advantage of improvements which had occurred in the interim.

GARY ROLIH COLLECTION

used to support the messenger wire wherever possible. This practice continued within tunnels, except on the insides of curves where a pin-type insulator was used on account of clearance. Nearly all fittings used in tunnels, including the messenger wire, were bronze in order to prevent corrosion. The hangers which attached to the messenger wire were made extra long, a result of experiences gained from the first installation.

Besides regular work and wire trains, a special concrete mixing train was placed into service to construct the foundations for the support towers. This was the leading section of the construction gang. It had been discovered that a self-contained unit train could carry on concreting with a minimum of disturbance to regular train movements. The train was composed of several special cars. A gasoline-powered mixer was mounted on a flat car. A water tank car was provided and materials were wheeled from adjoining cars along runways. It was possible to pour a six cubic yard foundation in 35 minutes from start to finish. In addition, a method was devised to heat concrete materials and water in cold weather. This allowed the crews to continue to pour concrete in temperatures as low as 30 degrees.

After a foundation was poured it was allowed to set for three to six days. The forms were then stripped and catenary tower erection proceeded. The supporting towers were assembled and bolted onto the foundations by a following tower gang. A work train using a crane would set the cross member to form the catenary bridge. The catenary tower was then bolted together and welded solid where required. Another following set-up gang would then bolt on necessary insulators and tower accessories. Hinged pulleys, called "snatch blocks," were attached to the insulators. A starting rope was pulled through the rollers running from tower to tower. To the end of the rope was attached the messenger wire. The slack was taken up in the rope and the wire pulled through the pulleys. The tension was taken up in the wire. A following work movement, called the wire train, was equipped with specially built tower cars and would clip-in the wire. This was accomplished by opening the hinge on the side of the pulley, pulling or "snatching" the wire over to the insulator, then clipping it in place. By this method, work progressed westward, with one gang following the other.

Two new substations were built at Roderfield and Iaeger to feed power from the high tension line for the extension. The design of these facilities differed from substations built on the initial electrification. These were constructed on concrete. The input power bus which led to the step-down transformers and the lead to the catenary feed were exposed to the elements, unlike prior installations that were housed in a structure, which had proven to be an unnecessary expense.

Other projects were undertaken to upgrade the facilities along the extension to facilitate operation under the wire. Work was begun in May of 1923 to enlarge the shop at Auville yard. Completion was targeted for the first of September. A portion of the old shop had to be dismantled in preparation for the erection of some of the new buildings. Several carloads of new machinery were received for installation in the shop. In addition to equipment associated with the electric locomotives, two Norfolk-type coaling stations were authorized to service steam powered mine shifters working out of the facility. The yard itself underwent an expansion at this time to accommodate increasing coal traffic originating on the Dry Fork line.

Improvements were made at the Bluestone power station. Already the facility was an impressive site. The powerhouse smokestack was the tallest on the N&W system. Two additional Stirling-type boilers were added, each with a heating surface of 6,800 sq. ft. The new total of fourteen boilers in the powerhouse had a capacity exceeding 8,000 H.P. Inside these boilers, 500 tons of coal was burned daily producing power for 600 wire miles of transmission lines. The station generators could produce enough electric current to light a 50-watt bulb in over one million homes. By the mid-twenties, the railroad was now moving over 15 million tons of coal yearly east over the electric lines to Norfolk and connections, averaging 41,564 tons daily. An additional generator was installed at Bluestone to aid with increasing capacity.

An extension was constructed to the storehouse at Bluestone Junction. By October 1923, the foundation for this building was nearly complete and the carpenters reported to begin work within a few days. The extension was authorized to store additional supplies and materials associated with the Class C-2 locomotives. A revision of the electric locomotive classification system took place on April 26, 1926 by authority of A. Kearny, Superintendent of Motive Power. Pending the order for new engines, the Class C locomotives were reclassified LC-1. The newly arriving power from Alco-Westinghouse would be assigned to Class LC-2.

Extensions to the cooling ponds for the Bluestone powerhouse water supply were authorized. John P. Pettyjohn & Company undertook the pool construction. Work was begun by first excavating for the expansion; then concrete work and piping began. The new cooling pond increased the capacity of the existing cooling pond by 50%, and was used with the turbine condenser system. The expansion was necessary since the river never provided quite enough water in it's normal flow for operation. About 800,000 gallons of water were used per day in the workings of the power station. After completion of the new cooling pond, the firm stayed on site a few weeks longer reinforcing the old cooling pond. Perhaps the most popular improvement, from the viewpoint of the operating people, was the installation of an American Water Softener and Filtering outfit. Water used for drinking purposes at Bluestone had always been very poor.

The site at Bluestone was a compromise from the beginning. Simply put, there was not sufficient year round flow in the Bluestone River without diverting a significant portion of the water. As a result, the effluent was more or less muddy and full of sediment. The softening plant removed the foreign material from the water by passing it through three feet of very fine sand in the filtration tank. The water then passed through another layer of coarse sand about eight inches thick. After passing through these two layers of sand, the water was forced by gravity and air pressure through a strainer. The water also passed through an alum tank, which had a lot to do with the clearing process and the improved quantity. The extension to Iaeger fully loaded the capacity of the power plant due to this limited supply of condensing water.

It also loaded the capacity of the transmission lines to Welch. The transmission lines for the Iaeger Extension were designed for a maximum voltage of 88,000, although initially it was set up for 44,000 volt operation. At some point in the future the railroad planned to step up the voltage from 44,000 to 88,000 at a transformer installation to be constructed at Welch.

Construction over the 22 route miles from Iaeger to Farm were reported to be progressing nicely in a report dated December 31, 1924. Expected completion date was April 1, 1925, which would bring a total of 60 route miles and 200 track miles under wire. Besides the addition of two electric locomotives, the ever increasing coal traffic required the purchase of 1,000 all-steel gondolas, 1,443 all-steel 57.5 ton hoppers and 6,000 all-steel 70-ton hoppers from commercial builders.

In 1925, the railroad exceeded a milestone, producing 100 million dollars in operating revenue for the first time. Operating ratio reached 64.57%, the lowest since 1917. A 45.6% increase in net income was attributed to the improved operating ratio and a significant increase in freight revenues. The road added 2 additional Class LC-2 electric locomotives, 4,000 new all-steel 57.5-ton gondolas and rebuilt an additional 2,057 coal cars in the existing fleet.

Examinations of the traffic producing this revenue revealed some important changes were taking place on the railroad. These would, in time, eliminate the need for further extension of electric operations. Coal destined for market on the west end of the railroad had continued to grow at a faster rate, than coal moving east. When the Ohio Extension was first opened, the industry and markets in the mid-west were still young. In 1901, coal moving eastward exceeded coal moving westward by a 4-to-1 ratio. As heavy industry began to expand in the Ohio, Indiana, Michigan, and Illinois markets, the demand for coal rose rapidly. During this period, eastward moving coal increased only slightly. By 1912 about the same amount of coal was moving east as west, which continued at this ratio throughout the World War.

In 1921, industrial development in the west began to greatly accelerate and westbound coal tonnage soared. That year, N&W moved slightly over 10 million tons westward, which increased to over 30 million tons by 1929. Eastbound tonnage remained flat, which meant that N&W's electric operation was not where the action was taking place. N&W's expenditures would offer a better return by easing transit to markets in the midwest. Expansion of electrification further westward was not necessary as it would not benefit coal traffic now moving in the opposite direction of electric service.

Movement of westbound coal did not involve the obstacle of heavy grades. Consequently the railroad could improve its profits by further encouraging the growth of westbound traffic as this business incurred a lower cost of transit. The railroad moved in this direction by both expanding facilities and opening new lines. In the mid-twenties, N&W changed the course of the Tug River in order to expand the classification yard at Williamson WV. At this location, coal was collected and classified for movement west. In 1931, the railroad completed rebuilding of its Buchanan branch, (from the narrow gauge Big Sandy & Cumberland), converting it into an efficient coal line. In 1933, a line was built across the mountain from Wharncliffe to Gilbert. N&W began moving a share of coal off the Virginian Railway to mid-west markets by this route. This traffic all moved behind steam motive power.

The extension of the electric lines from Farm to Iaeger, including substations at Roderfield and Iaeger, and the extension of the branch into Auville yard was finished and placed into service in 1925. Double tracking on 1.67 miles of the Tug Fork branch was completed, calling for electric catenary to be installed on this track.

An additional 1.58 miles of catenary was placed into operation to reach Clift yard, a coal train assembly point on the Bluestone Branch. Finally, an expansion of this yard to accommodate growth in coal loading required 2.05 miles of new electrified lines to be installed over tracks in this facility. Construction was begun for a high-tension line between Bluestone powerhouse, running up the Bluestone branch to Matoka. This was scheduled for completion in 1926, and would allow for the interchange of electric current with the Virginian Railway, a safety measure to back up operations.

Also moving ahead was construction for foundations for catenary extension to Williamson, as the final decision to cease westward expansion of the electric lines still had not been made. As early as 1916, the railroad had anticipated electrification to Williamson. Working towards this goal, N&W had engaged Gibbs & Hill to determine the economic benefit of such a move. The full impact of mining further to the west of the Elkhorn, which had not matured in 1916 was not taken into account. The report stated return on investment from electric transport of Elkhorn coal at that time would be a minimal 5%. The study further indicated additional expenditures would be necessary if a Williamson Extension was undertaken, the largest proportion of this being a new power plant. It was recognized then that the Bluestone plant lacked adequate water supply for further expansion and the report selected a site at Thacker, WV.

The railroad had additional plans, which in part would relieve the issue of a new electric power supply, filing with the I.C.C. on July 2, 1925 to lease the Virginian Railway. N&W President Maher had administered the Virginian under the USRA authority and had seen first hand the benefit of co-ordinated services between the two lines. Control of the Virginian had been a contested issue since the end of World War I, at which time that road's coal business had matured and become profitable. Further interest in a Virginian merger had been spurred by the Transportation Act of 1920, which directed the I.C.C. to prepare and adopt a plan of consolidation of railroads into a limited number of systems. Its purpose was to solve the problem of a number of weak roads, and establish an equal base from which uniform rates could be established. In practicality, it stirred up a number of controversial issues and conflicting merger schemes.

In the Pocahontas region, various plans were proposed but the issue of the Virginian was soon contested by Norfolk & Western Railway and the Chesapeake & Ohio. Neither road wanted the Virginian to fall into the rival's camp. In 1921, N&W had tried to work out an agreement for a Virginian lease but this effort failed because of an excessive number of accumulated dividends. In 1923, C&O proposed either a joint lease of the Virginian with N&W, or purchase by C&O alone. This was rejected by the Virginian, which regarded C&O's credit to be inferior to that of N&W.

On July 17, 1923 a truce was signed by both C&O and N&W, and later extended until January 16, 1924, in which all agreed that neither road would try to lease or control Virginian. At this time, negotiations reopened between the two for joint control. From the C&O point of view, the Virginian had been an invasion of its territory; from N&W's point of view, joint control would give C&O an entrance into Norfolk. In addition, N&W, through it's Pocahontas Coal and Coke Company, had acquired 120,000 acres of coal land in the Guyandot Valley and wanted to build rail lines into the region. Virginian and C&O already operated in the Guyandot region, and who would move westbound coal from this area also became part of the controversy.

In February, 1925 N&W began negotiating with representatives of the Roger's estate, which owned the bulk of Virginian stock. Soon they came to acceptable terms. On April 22, 1925 negotiations were made public and the following month were approved by stockholders of both roads. C&O soon opposed the move, which began an organized effort to defeat the proposal, which included filing to construct a number of branches into the Guyandot area. The fight and the terms of the lease were carried before the Interstate Commerce Commission. The Norfolk & Western presented the terms of its 999 year lease in which it agreed to meet all charges, taxes and expenses, pay all funded maturities, and guarantee a 6% return of stock. The railroad was relying on a number of financial and operating economies to make the lease possible. Among these were certain changes to the electric operations.

In testimony before the I.C.C. on October 12, 1925, N&W's new President, W. J. Jenks, stated that 10 million dollars could be saved by utilizing Virginian's electric power station at Narrows VA, and reassigning its locomotive fleet. Instead of spending funds to construct a new power plant at Thacker, WV for the extension to Williamson, Virginian's station at Narrows could be utilized to provide power. Some Virginian equipment had been taken directly from N&W designs, while the power distribution system incorporated improvements suggested from Gibbs & Hill, consultants for both roads. The Virginian system was designed to deliver extremely large amounts of power necessary for heavy tonnage and high speed operations over longer distances. By combining the two systems and extending electrification west on N&W, this would release 36 Mallet-type locomotives, worth 2 millions dollars. Other savings included better movement of coal cars, economies gained by new connections between the roads, and the elimination of the need to double track the Virginian at some point in the future.

A large amount of opposition was mounted against the merger application, with C&O at the forefront. It attempted to dispute the operating economies, particularly savings from double-tracking the Virginian which C&O argued would never become necessary. Joining in opposition were representatives from towns along the Virginian who felt they would be hurt by lessened competition. Another protestor, the Logan County Coal Operator's Association, feared the acquisition would mean the loss of their eastern outlets. The I.C.C. weighed the arguments and concluded the lease would eliminate existing routes and disapproved the merger.

Mr. W. J. Jenks served as president of N&W during the attempted lease of the Virginian Railway. Had the lease proceeded, electrification was promised to quickly extend westward to Williamson.

An expansion of Norfolk & Western's electrification and the possibility of unified operation of the two road's electric zones came to a conclusion. Traffic flow had changed on N&W with more coal now moving west. The railroad concentrated its resources in supplying coal for the rapidly growing industrial base in the midwest. Concrete foundations for catenary extension along the N&W mainline between Iaeger and Williamson were completed in 1926. The decision to proceed with the Williamson Extension was delayed while first labor unrest, then depression visited the coal fields.

Virginian opened the initial 14-mile section of its electrified line between Elmore and Clark's Gap in September, 1925. By the following year, Virginian electric locomotives were running all the way to Roanoke. An integration of sorts took place on a smaller scale when an interchange was developed with Virginian Railway at Matoka, WV in 1925, to give that railroad another access westward over N&W to the Great Lakes and northwest markets. Electrification passed from its period of expansion into that of a mature operation, doing best what was intended, the economical movement of coal. ■

5 Working on the Railroad

LC-2 2514 rolls a coal train eastward near Falls Mills, VA in 1930.

The Norfolk & Western Railway had become foremost, among other things, a coal hauling railroad. In obtaining this goal it led the industry in a number of pioneering improvements in addition to electric train service. N&W had been the first railroad to adopt the mallet-type locomotive for general road service. Before this success, contemporary rail operations had used these engines as helpers over short distances. The railroad's prodding of builders for increased evaporating capacity, and its internal innovations had a large part in turning this into a viable over-the-road locomotive. It was also the first railroad to use 100-ton capacity cars, nicknamed "battleships," in coal service. The gondolas were easily loaded from chutes at the mine breakers and at Tidewater the car could be put into a frame and capsized bodily. Coal was dumped into bunkers located in a new pier, constructed by N&W in 1913. This was the largest coal pier in service on the Atlantic seaboard, and capable of loading a modern 5000-ton cargo steamer in two hours.

Approximately 71% of all N&W revenues were generated by coal business originating along the railroad between Graham, VA and Williamson, WV. Within this 104 mile district the company held a monopoly on all traffic. About 2000 carloads of coal and 100 cars of coke were produced within this area daily. Before the start of electric train service, about 250 carloads a day went westward, principally through Columbus, Ohio for rail distribution throughout the midwest. The balance went east to Tidewater or connecting lines.

At the beginning of electric operations, the most

concentrated mining in the Pocahontas field took place between Graham and Welch, a distance of about 27 miles. There were 86 mines located in this section, producing 62% of the total coal tonnage of the whole road. It was calculated that over 5 billion tons of recoverable coal or 100 years of mining remained at then contemporary levels of activity. In addition, heavy merchandise traffic featuring five time freights passed through the region. With the exception of the steep Elkhorn grade, the remainder of the line, extending eastward from the coal field to Norfolk, had a ruling grade of 0.5%. West from the coal field to Columbus the ruling grade against coal traffic was 0.3%. However, the portion of traffic going eastward still had to contend with the steep Elkhorn grade.

Prior to electrification N&W's highly developed steam operation handled this coal traffic. Empties for reloading would be delivered from Tidewater into the receiving yard at Bluefield, WV, headquarters for the Pocahontas Division. At this point they were handed over into the outbound yard, and in the process inspected, classified and reassembled into trains for the coal fields in trains of 75 to 100 cars. By 1914, there were additional trains intermingled with the daily 10 to 12 empty delivery and slow coal trains working through the Elkhorn region, as the through business of the company began to mature. The time necessary to clear 4 through passenger trains, 8 local passenger trains, and 8 time freights had to be taken into consideration. It was a difficult task to avoid delay to time

LC-1 2508 climbs the Elkhorn grade near Maybuery, WV. A general routine had each motor set in road service making two round trips per crew each day, an improvement over steam operations.

sensitive trains, yet keep the coal moving.

The handling of empties out of the yard at Bluefield, delivery to mines involving the picking up of eastbound loads, and normal steam delays for coal and water stops were a source of problems. The time lost in clearing for passenger and time freight movements further slowed coal movements. Additional delays at Elkhorn tunnel caused an average round trip of 11 hours for a double-headed steam road crew. This gave a total engine hours per round trip of 22 (for the engineer and firemen on each locomotive) chargeable to the crew out of Bluefield to which was added about 4 crew hours for the pusher, or about 26 crew hours to move one coal train under steam operation under ideal conditions.

To further assist operations in the preceding 18 to 20 years, the railroad's general policy had been to reduce operating expenses by new construction, adding double track where necessary, and relocating the line to eliminate grades and curves whenever practical. Both increasing numbers and capacity of rolling stock were balanced with improvements to motive power and roadway conditions. This balanced planning allowed maximum benefit to be obtained from any improvement involved in the handling of heavy tonnage trains. Still, the increasing amount of system-wide business when taken into consideration with the difficulties of operating conditions on the Elkhorn grade called for a quick solution to the choke-point on the railroad.

The first electric operation started February 10, 1915. Electric and steam operation intermingled until April 14, 1915, when electric locomotives were able to assume all eastbound coal movements. With the use of electric locomotives, speed on the Elkhorn grade increased from 7 to 14 m.p.h. Besides the faster transit time, the ability to "turn" an electric engine (that is, make it available for service again) was considerably quicker than steam. With mallets, this was a time-consuming process, requiring 10 to 12 hours to recondition a locomotive after a run. Not only was there much work to be done, such as cleaning the fire, but often the Mallets had to wait their turn.

The first LC-1, 2508, assists train No. 4. The Pocahontas, *upgrade at Switchback, WV in 1929. Electric power became more important in cold weather as the thermal efficiency of steam declined. The electrics routinely assisted on all heavy trains.*

Naturally, engine crews preferred to operate the electric engines that were, in comparison, clean, easy to operate and reliable. Electric engine crews had a more regular work schedule. Coal traffic was run as the business warranted, with crews called on 8 hours notice. The men could count on being home at certain times. The crewmen assigned to electric engines were required to be qualified for both steam and electric operation. Steam experience posed no problem as the electric jobs were in demand and usually filled by men high on the seniority list.

Steam engineers usually qualified for electric runs after about a week of instructions given by an assistant road foreman of engines. Instead of firemen, the electric locomotives had "helpers" who also underwent special training. The instructor was familiar with every detail of electric train equipment. The training began in the shop, where the different parts of the equipment and its function were explained. Emphasis was placed on procedures to be followed in case of an accident. The instructor then went into handling of trains in regular service and the engineer's practice in starting up and shutting down an engine. Following this portion of the training, an examination was given, and successful completion certified the men for electric operation. In the initial runs of new engine men, the railroad paired them with experienced helpers. In this manner they were not bothered with details that the helpers could handle.

The assistant electric road foreman of engines also held a meeting once a week, (usually Monday mornings), open to all who desired to attend, during which all facets of electric operation were covered. Steam engine personnel often attended these meetings in order to prepare for electric runs without interference with their regular assignments. Also, special conditions existed for steam crews operating in the electric region.

In the Elkhorn region, where the two types of motive power intermingled, special safety precautions had to be taken. These were covered in the meetings and rigidly adhered to for obvious reasons. The most important rule for electric trains was what to do when the power failed. Under those conditions, the training specified that the train was to be stopped and hand brakes set. This was done to avoid the chance of damage due to air failure. Steam locomotive personnel also had special duties for which they received training in areas where the two types of power intermingled.

Before entering electrified territory, crews inspected steam locomotives so that no parts of the engine or engine appliances, such as cab ventilators, firing tools, or coal on the tender could foul overhead wires. Further instructions specified that firing tools would be used in such a manner that they could not come within 3 feet of the overhead wires. The squirt hose stream was not allowed to strike overhead electric wires.

All employees were prohibited from going out on running boards or smoke box steps without first secur-

Electric engines took over coal train duties from steam, which remained in use on passenger and merchandise trains passing through the region. The LC-1 electric locomotive powering this coal train east from Falls Mills sports both a high and low mounted headlight.

ing permission from the person in charge, (conductor or yard master), who was to supervise the movement of employees while in such position. All employees who were required to go on running boards or smokebox steps while the steam engines were under overhead wires could not go within 8 feet of transmission wires. Employees were forbidden by timetable restriction from getting within 18 inches of live catenary or signal transmission wires. Further instructions concerning hand signals given from the tops of cars directed they not be done under the trolley wire. Attention was called to the fact that wires could be lower than the standard 23 feet, 9 inches under or near bridges, and in or near tunnels. Whenever possible, engines were not stopped with the stack under porcelain or wood stick insulators, so as to prevent damage to these appliances.

Steam locomotives could only take on water, coal or sand at specified locations while under the contact wire . While steam engines were standing on electrified tracks, employees were not permitted on top of the tender until the standpipe was moved into position for taking water. The ladder on the rear of the tender was not used in the electrified zone for ascending or descending from the top of the tender without taking care to note the location of the trolley. Employees were not allowed to descend from the top of the tender before the standpipe was returned to normal position.

At certain tight locations in coal assembly yards, water standpipes were equipped to permit de-energizing of the contact wire for at least fifty feet in each direction. This was done automatically by a knife switch when the standpipe was moved into position for taking water. When electric helper locomotives were coupled to assist a steam engine requiring water, the pantograph of the electric engine was moved to the down position. This was done before the standpipe was moved from its normal position, to prevent bridging from an energized wire to a grounded trolley section. Pantographs were not to be raised until after watering activities were complete and the standpipe was returned to normal position. Electric engines operating on a track which was occupied by a steam engine were not allowed to pass a water limits sign until after the standpipe was returned.

Fourteen miles an hour was calculated as the maximum safe speed at which the electric coal trains could be safely handled through the Elkhorn field. This was done with respect for the right of way, which was laid out in a comparatively narrow valley, and was encroached on either side by buildings, coal tipples and coke ovens. Over 60% of the railroad was located on curves, with 54 localities for switching movements between the Eckman assembly yard and the Elkhorn tunnel.

The three-phase electric locomotive permitted the railroad to maintain a constant speed of 14 m.p.h. while moving upgrade, then, once over the summit, switch to regenerative braking when moving down the east side of Flat Top Mountain at a maximum speed of 28 m.p.h. This allowed telegraph operators, tower men and dispatchers to closely figure the times on these tonnage movements. It became a daily practice to handle these trains directly ahead of local passenger and time freight trains. In these cases, the tonnage coal trains were given a five-minute time margin with which to clear the following train at the single track Elkhorn tunnel. When operating at 14 m.p.h. an electric coal train could clear the tunnel in approximately three minutes. This was a considerable improvement over steam operations, especially when combined with the total absence of stalling coal trains.

The Coaldale tower operator and telegraph signal CQ, controlling movement through Elkhorn tunnel, could figure the time on the electric trains after they struck the track indicators to within one or two minutes running time. The interlocking plant located at Coaldale operated switches located on both sides of the tunnel. The tower was equipped with a 17-lever machine with 13 working levers that operated one 3-arm, four 2-arm, six 1-arm and 3 dwarf signals. Along with five derails, these controlled a total of eight switches.

The electro-pneumatic controls were operated by air compressors that had a capacity of 50 cubic feet per minute. The compressor was powered by a 15-h.p. single phase, 60-cycle electric motor. Instead of extending an air line through the full length of the plant, two additional 15-cu. ft. per minute air compressors were installed to work a crossover controlled from Coaldale that was 5,500 feet further westward down the hill. These compressors drew power directly from the transmission line. With the certainty of train movement gained from electric operation, the tower operator could permit an eastbound or westbound train the right-of-track without any preference or other interruption to traffic. This ability eliminated Elkhorn tunnel as a bottleneck.

Further benefits were soon discovered in electric operation. During zero degree weather these engines operated at practically full tonnage rating, as compared with steam engines which required a 10% tonnage reduction. In fact, the efficiency of the electric engines increased as the temperature decreased. The equipment required for electric locomotive maintenance was considerably less than needed to take care of steam engines. The absence of coal wharves, spark pits, water tanks, roundhouses and turntables, incident to the handling of steam, was a considerable economy.

Coaldale Tower controlled train movements through the single-track Elkhorn Tunnel with a 17-lever machine, working a total of 8 switches.

This postcard view from over the top of Elkhorn Tunnel looks at the grade approaching Coaldale from the west. The operator at CQ Tower controlled the crossovers in the distance which were utilized to keep the traffic moving through the single track tunnel without delay.
Roger Nutting Collection

By summer, a general routine of handling engines and crews had been established. One motor set with its engine crew made two round trips per day. Normally a crew was called at Bluefield and took a train of empties westward. Once through Elkhorn tunnel, they would set off as required until reaching the coal assembly yard. They would then turn to make their first trip eastward with 3250 tons to Flat Top Yard. Here, they would set off the train and return to Eckman, where they would pick up another 3250 ton train. The crew would then work eastward to Flat Top Yard where they would pick up approximately half the tonnage brought over on the first run and continue eastward to Bluefield.

On some occasions on the first trip into Flat Top, sufficient tonnage was on hand to allow a crew to fill out to a maximum of 4700 tons, then continue east to Bluefield. From Bluefield the crew would turn westward, usually with another empty hopper train.

Besides coal loads brought over the hill from the Elkhorn, tonnage continued to flow from mines into Flat Top Yard off the Bluestone and Pocahontas branches, some of which was assigned for westbound movement. Coal was collected from the mines by steam shifters on the Bluestone branch and assembled into trains at Clift Yard. Electric engines operating from this point brought the loads upstream from Clift Yard through Bluestone Junction, and three miles further eastward to Flat Top Yard. Clift Yard was a 5-track facility, oriented south to north, stretching over 1.5 miles along the Bluestone River. The west (north) end of the yard laid near MP B-6, with the tracks twisting south through a series of 8 and 14 degree compound curves. The yard office and storage track where electric power would lay over were situated near the east (south) end.

Branch line rail was 130 lb. installed to accommodate the heavy coal tonnage. Trains feeding into the mainline at Bluestone Junction departed Clift yard on a nearly nonexistent 0.05% grade which stiffened to 0.27% at MP B-3 north of Bramwell. Curvature stiffened again to 14 degrees near MP B-2 as the rails passed under mainline Bridge No. 854. The branch line mileposts at this point changed to reflect the line's former mainline status passing under the bridge. At MP N-375 at Cooper, the grade on this line increased to 0.5% which was the ruling grade as the right-of-way followed the compound curvature eastward to Bluestone Junction. Once out onto the main, a three-mile sprint eastward brought the trains into Flat Top Yard. At this facility eastbound cars were added to other coal loads waiting movement at Flat Top, while westbound loads were set out for pickup by westbound trains.

Coal coming off the Pocahontas branch was also moved by electric traction. A single locomotive set performed normal operation of this branch line making a number of trips. The Pocahontas branch left the mainline at SU tower, immediately crossing Bridge 852 on a 5 degree curve. The grade stiffened to 0.5% as the line worked into a 9 degree curve, under a Low Wire Warning for the catenary and into 201 ft. Pocahontas Tunnel No. 1. The grade increased to 0.7% along with the curvature which continued into 366 ft. Pocahontas Tunnel No. 2. The VA/WV state line was crossed inside Pocahontas No. 2 Tunnel. Passing MP B-1 the grade increased to 0.75% just east of the Pocahontas Yard board. This facility was laid out on curvature

Coal awaits the journey eastward in the assembly yard at Flat Top, VA as seen from the cab of a passing steam run. Although coal traffic moved behind electrics, steam motive power continued to operate through the region on both merchandise and passenger trains.

ranging from 4 to 15 degrees on the narrow valley floor opposite the town. An usually large number of trains moved out of this assembly yard through Bluestone Junction to fill out road trains at Flat Top Yard.

In pusher duty, an electric locomotive made five trips shoving eastbound coal trains over the Elkhorn grade in a normal working day. The pusher would begin it's day with the crew registering in at Bluestone Junction. The crew would work west following a hopper train to assist with setting off cars for coal loading. The helper would work west to a coal assembly yard, before it turned to assist a loaded coal train up the hill and through Elkhorn tunnel to Ruth. Here with the help of a hook it uncoupled on the fly and returned west to push another coal train. As with steam operation, the Bluefield crew continued to distribute empties on the westbound trips, assisted by pushers returning downgrade. Full tonnage eastbound coal trains were handled by two motor sets with one on the head end and one pushing. Heavy manifest freight trains used one motor set pushing to assist the steam road engine. A heavy passenger train requiring assistance would get one motor helper assisting on the head end.

A report on train sheets during August 1915 showed some of the variations that occurred in the normal level of activity. On August 1st, eight crews out of Bluefield made 16 eastbound trips with 581 loads, no empties, for a total of 51,501 tons. Westbound, these same crews delivered 361 empty cars and picked up 150 loads. The Elkhorn pushers, with four crews, made a total of 18 helper trips. Another crew, known as the Flat Top pusher, made 8 helper trips. This crew largely assisted empty hopper trains in working mines, as switching was an important part of the duties of electric operations.

On August 9, 1915, the eight crews out of Bluefield made 16 trips, bringing 51,849 tons over the mountain. On the outbound trip they dropped 320 empties and picked up 109 loads. The Elkhorn pushers with 4 crews made 19 assists, with the crew on the Flat Top pusher making 15 shorter trips near the top of the hill. Later in the month, on August 19, 1915, the outbound crews dropped 744 empties, picked up 171 loads and returned eastbound with 59,291 tons. The four crews working the Elkhorn pushers made 20 trips while the Flat Top pusher made 6 assists.

The following day, August 20th, was an extremely light day. Three crews out of Bluefield dropped 427 empties, picking up 33 loads. Eastbound they returned with 149 loads, and 126 empties for a total tonnage of 15,857. The four Elkhorn pusher crews made 16 trips and the Flat Top crew made 6 trips.

During winter weather, electric engines continued to operate at nearly full tonnage. A string of electric locomotives lays over in the yards at Bluefield, WV.

The last unit of Class LC-1 2511 pokes through a coal assembly yard. The heat generated by the locomotives could become unbearable during the summer months.

The electric fleet was dispatched from the motor pit at Bluestone Junction. This was the headquarters of the entire electrified division and center of operations. The inspection building and machine shop for electric locomotive maintenance were located here in order to be in close contact with other parts of the operation. The power director was in contact with all points, including train dispatchers, yardmasters and superintendent, and used these contacts to regulate locomotive use. An interlocking tower was located at Bluestone Junction and controlled train movement through the area. This plant had 22 working and 2 spare mechanical levers, in addition to 14 electric levers. Approach and route locking was provided for all through movements which, as detailed in a featured article in the N&W employee magazine, averaged 150 movements during a 24 hour period.

All electric locomotives in service went over the inspection pit at Bluestone Junction once each day. This inspection took about 40 minutes in a building located next to the powerhouse. The motor workforce was divided into different areas of specialty, in which 5 to 6 men worked on each engine as it moved through the building. The machine shop used for locomotive repair also doubled as a location to perform any needed work for the powerhouse. This inspection building and repair shop was a 148-ft. by 68-ft. brick structure similar in construction to the powerhouse located just to the north.

The locomotives were not run into the building under their own power, but were brought in by means of a cable and motor driven winch. This winch was located at the east end of the building between the two main tracks. There were two locomotive pits running the length of the building. At the west end of the shop was a short wheel pit, connected to the two other pits by a covered lateral pit. This pit was equipped with a narrow gauge track to accommodate movement of a pneumatic jack. By means of this, the wheels were removed from the locomotives and run along to the storage track in front of the wheel lathe. A blacksmith shop was located in an extension at the northwest corner of the building. The southern pit in the inspection building was used exclusively for locomotive inspection.

To assist, a 112-ft. long platform was located alongside, at 12-ft. 8-in. above the rails. All repair work was done in the northern part of the building where the machine tools were located. This part of the shop, as well as one of the inspection pits and the short wheel pit, were all spanned by a 30-ton Whiting crane. The machine shop was equipped with an 80-in motor driven heavy wheel lathe, a 30-in by 8-ft. planer, a 60-in. by 6-ft No.4 horizontal boring and drilling machine, a

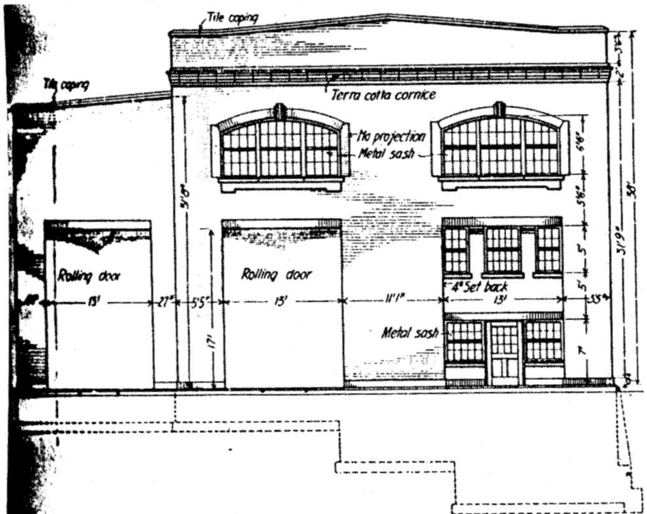

East Elevation of Inspection Building for Maintenance of Electric Locomotives, Elkhorn Division, Norfolk & Western Ry.

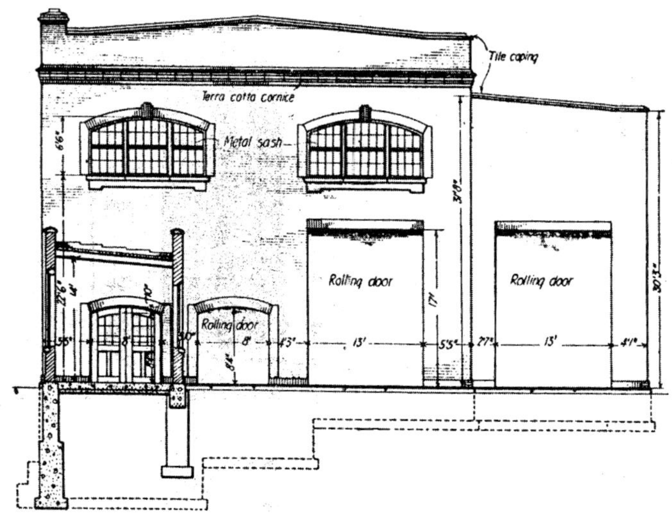

West Elevation of Inspection Building for Maintenance of Electric Locomotives, Elkhorn Division, Norfolk & Western Ry.

The east and west elevation drawings of the maintenance shop at Bluestone Junction gives and understanding of the size of N&W's facilities.

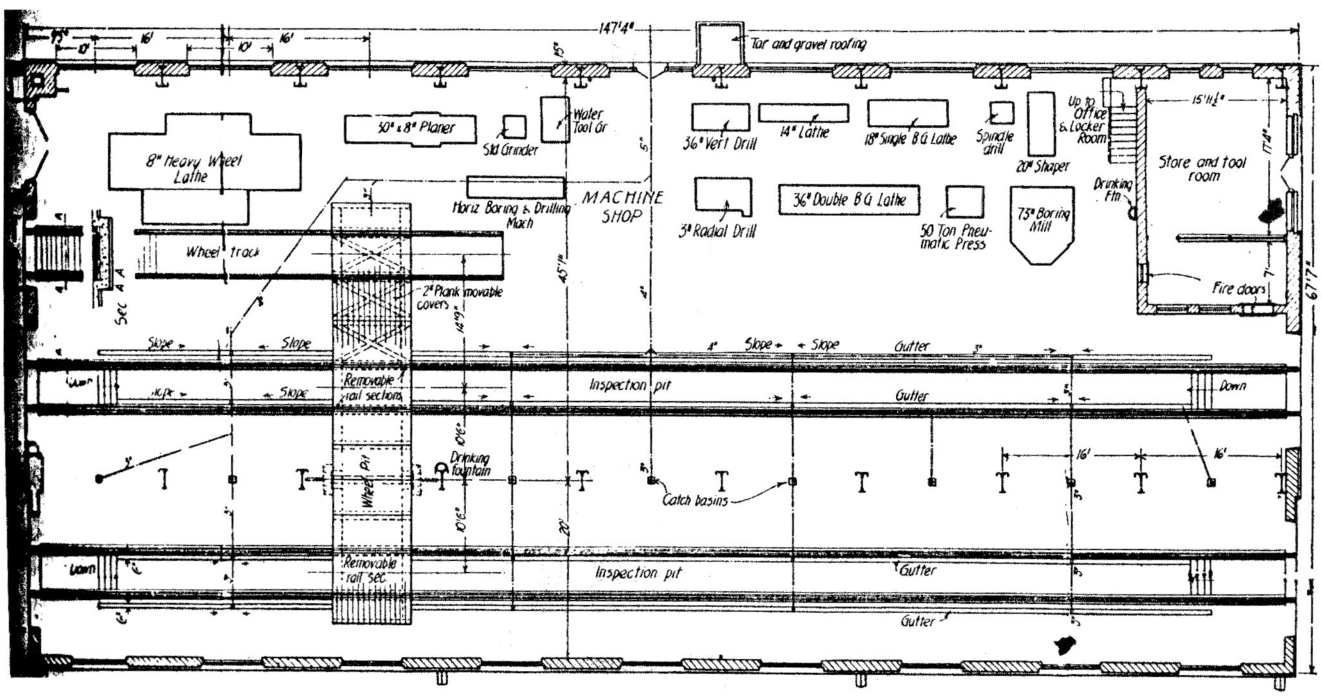

The floor plan of the building shows the loation of equipment used in the care and repair of the locomotive fleet.

Both Drawings from Mason Y. Cooper Collection (Originally published in Railway Age.)

3-ft Mueller motor driven radial drill, a 20-in. Stockbridge motor driven crank shaper, a 73-in motor driven boring mill, and a 36-in. double back motor geared lathe. In addition the shop was outfitted with a number of belt driven tools, including a Style C Diamond double emery grinder, a No.5 mat diamond wet tool grinder, a 36-in. Aurora vertical drill, an 18-in. by 10-ft. single back geared lathe, a 14-in. Barr 2 spindle sensitive drill, and a 50-ton pneumatic press. The building was equipped with electric Tungsten lighting. Rolling steel doors were provided at the entrance to all tracks, which allowed the building to be steam heated from the powerhouse.

Further changes came to the shops in the early 1920s when the railroad ordered an additional four pairs of locomotives from Westinghouse Alco. They were similar in operation but had nearly a third greater capacity than the Class C (later the LC). With an hourly rating of 4750 H.P. and a maximum tractive effort of 168,000 lbs., they quickly became favorites among crews. Tonnage rating on the Elkhorn increased from 3250 tons behind the Class Cs to 4000. The C-2s fell into most of the road train duties, with the Class Cs filling in the remainder as well as handling pusher duties. The mechanical gear varied somewhat, calling for additional parts inventory to be carried at Bluestone Junction. The C-2s used a rigid 2-8-2 or 1-D-1 wheel and spring arrangement that was close in operation to a Mikado-type steam locomotive. A larger single Westinghouse model 452 inductive motor was geared to each jack shaft requiring its own special parts and tools.

In 1924 several changes were initiated at Bluestone Junction in order to make the job easier for the men repairing the Class C-2 locomotives. These engines were much heavier, and the drop pits were reinforced with new timber and concrete. A new storage building, of wood frame construction and covered with galvanized iron was built on a concrete pad at the east end of the locomotive inspection bay. With completion of the structure, additional space for the tool or brake rooms and new machinery was available in the shop. This space was designated specifically for the C-2s. Congestion within the inspection building was relieved, by removal of all spare heavy electrical and mechanical repair parts to the storage building.

Once each day, every electric motor in service was rotated over the inspection pit. Inspection by shop men involved the usual mechanical adjustments, taking up any slack which may work into the brake rigging, inspection of electrical contactor for burns and pitting as well as other details of normal wear. Once a week the engines were rotated through the general shop for dropping out of wheels. With the excessive curvature of the Pocahontas Division, wheel flange inspection for wear was a constant source of concern. If a wheel change-out was necessary, this job took about a week to complete. The individual motor cabs were interchangeable, so that in the case of trouble with one unit, pairs could be mixed to keep the maximum number of locomotives in operation.

After the arrival of the C-2 engines, the capacity to handle coal increased. To assist in this work a fourth generator was added to the Bluestone Junction power station in the space previously set-aside for this purpose. Eastbound coal moving under electrification averaged over 40,000 tons per day. Sixteen locomotives were now available for routine service, with tonnage per train now exceeding 3700 tons. With the improved reliability of the new locomotives, the Bluestone Junction motor pit began making and breaking all records in 1923. In April, 295 dispatches of motive

LC-2 2512 rolls eastward near Bluestone, WV in 1930. The arrival of the LC-2s allowed the capacity of the line to increase without additional trackage by handling more tonnage per train.

power were made with only 37 minutes delay. In May 1923, 270 dispatches were made with 63 minutes delay, 49 due to equipment and 14 waiting on crews. This record was not matched anywhere on the road in steam operation.

Much notice was paid by the industry to N&W's pioneering use of electric regenerative control, which by 1924 had been in routine operation for nearly ten years. In early days of electric operation, attempts had been made to brake jointly with air and motor. This resulted in very heavy shocks to the train, broken couplers and separations. The practice was soon abandoned. Regenerative braking had since settled into routine use, and was now touted as an advantage of electrification. The railroad hosted visiting delegations, in part to observe regenerative control in operation both from the cab and the powerhouse.

Riding in the cab was a different experience for these delegates which may have been more accustomed to steam locomotives. The sensation of the view from the cab was more like a streetcar than a locomotive. Close attention was paid to the operation of the controls. As the locomotive passed over the summit at the east end of Elkhorn tunnel, the engine man closely watched the ammeter. Selecting a time the train was balanced over the summit was a learned skill. There was a time interval needed to change the phase converter connections. The possibility of a runaway occurring within this period was real. The rate of acceleration on grades could be so rapid that the wheels would slide back to synchronous speed, at which point regenerative braking control would be impossible. The engineer had to maintain a constant speed while making the change in order to prevent rough slack action. When the current began to drop, he shut off the controller and threw the balancing lever from motoring to regenerating position. He then reopened the controller to running position.

Regenerative control was similar in many respects to dynamic braking, a process later developed with the diesel-electric locomotive. With dynamic braking, the excess amperage was dissipated as heat. With regenera-

Regenerative control was used on the eastern side of Flat Top Mountain. Class LC-2 engines work a coal train eastward approaching Bluefield, VA on June 12, 1930.

tive control, the electricity was put back into the catenary for potential use on other parts of the system. Regenerative control allowed the N&W to handle long and heavy coal trains very smoothly, as 90 car trains could be brought to a stop without any jarring. When in regeneration on the heavy grade, and if for any reason the brakes had to be used, the rules obligated the crew to brake all the way to the bottom by air alone. Later this procedure was amended by timetable to include the use of retainers. These instructions were very specific in detail and required a period of time to fulfill. As a result, train crews used regenerative braking whenever possible to get over the road quickly.

Unfortunately, mishaps did occur in electric train service for a variety of reasons, resulting in rule additions. Perhaps the most unforeseen was the difficulty of the former steam engines crews in staying awake. These men had been accustomed to the constant activity and noise associated with keeping a steam engine in service. With the electric motors, the constant speed, both uphill and down, the droning sound of the motors, the warmth in the cab and absence of duties requiring close attention, put the most vigilant men in a somnolent condition. Once a train had gotten underway, there was simply nothing to do except watch for signals. Shortly after electrification, a head-on collision occurred at night between two trains on the Elkhorn grade, because a crew on a eastbound coal train had fallen asleep and run past a stop signal. The opposing hopper delivery crew, coming down the hill, saw the men asleep in the cab, and sounded an alarm in an attempt to awaken them. The collision occurred when they could not alert the crew or back their train uphill out of the way.

During early winter operations, sleet conditions caused a buildup of ice on the trolley. This caused arcing as the pantograph rode onto the ice, or was forced downward by the additional weight on the pantograph, resulting in the trolley wire burning through. Two engine men went on top of the motor against regulation, and were electrocuted when they attempted to clear ice. This led to the development of special sleet instructions. At first indication of sleet, a report was made to the Power Director or dispatcher by company telephone line, who would issue alerts to all engine crews. Engines were instructed to run with all pantographs up, in order to remove ice buildup, except through phase breaks. In cases of heavy sleet conditions locomotives were assigned to patrol train service, with the control switch on the motor cutouts open on the lead engine. In this way, the lead pantograph cut the sleet by abrasion. Electrified sidings and yard tracks were given attention with patrol trains. The wire train was manned at Bluestone Junction and made ready for movement during sleet storms in order to repair any wire damage.

Electric locomotives standing in yards or at the powerhouse had pantographs lowered and raised frequently during sleeting to prevent the accumulation of ice. During sleet storms, special attention was paid to the pantograph shoes on motors coming in for running inspections at Bluestone Junction. Further instructions were issued to crews on moving trains for sleet conditions. When the sleet load on a pantograph became excessive, the master control switch for the locomotive was shut off. An attempt was then made to raise and lower the pantograph several times. If this failed, the train was stopped under a contact wire that was 5 or more feet above the roof of the engine. An additional attempt was then made to raise and lower the pantograph several times using the pantograph pole, paying special attention to avoid contact with the overhead wire.

If the accumulation of sleet was such that the pantograph still could not be made to operate, trainmen were allowed to go onto the roof of the locomotive with special instructions that required strict adherence. The pantograph "DOWN" button was set in the "IN" position and all pantographs lowered. Ice was removed from step treads leading to the roof to insure safe footing. All grounding switches were closed and locking pins inserted. The position of all overhead wires were noted. Latches held down pantographs, and grounding switches had to make proper contact.

When working on the roof of the locomotive, extra precautions were taken against slipping. Employees were instructed to keep at least three feet from, and not work directly under, the wires. Standing astride the pantograph tubing so that it could not raise, the employees were instructed to lightly tap tubing, springs, rods, shoes and other parts to remove sleet. Excessive weight was not to be applied to the pantograph tubing which might throw it out of position. Before descending from the roof attention was paid to make sure latches held down all pantographs, and grounding switches were still closed. After all personnel had descended grounding switches were opened, with employees paying attention to keep as far as possible from all normally energized parts. The engine was then prepared for service. By using these procedures and safety precautions, no further injuries or electrocutions occurred. To further reduce the necessity for working on the roof, locomotive headlights were changed to a new mounting position located below the cab. The work was performed at Bluestone Junction during routine locomotive shopping. The amount of illumination differed at the low mounting, with some units retaining both high and low mounted headlights for a peri-

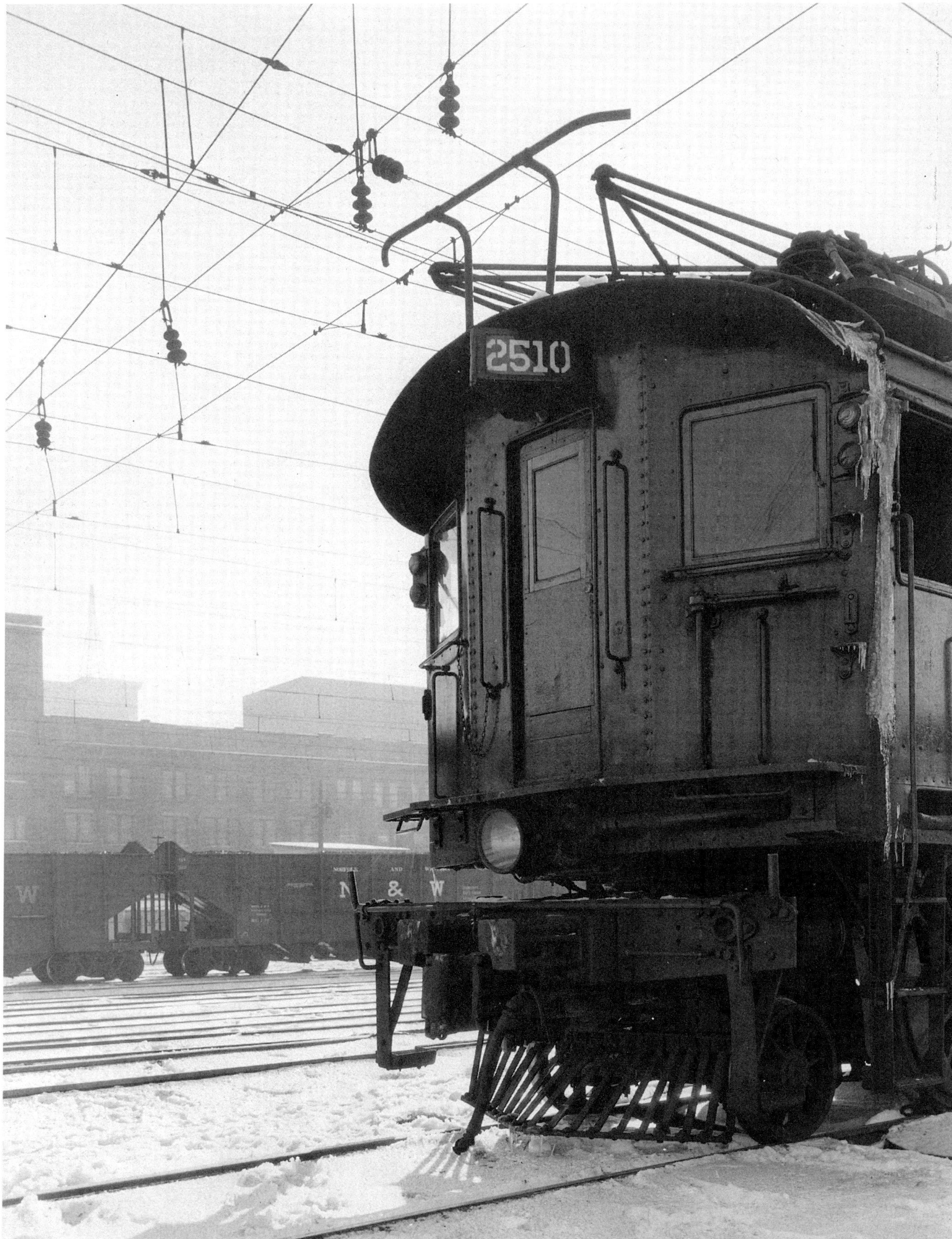

LC-1 2510 lays over in the snow at Bluefield, WV. Operations under sleet conditions posed hazards of arcing, which could damage the catenary.

od of time.

Trouble at the Bluestone powerhouse in May 1919 brought about a severe interruption in electric service. These failures were due to weaknesses in various components that were solved and replaced by Westinghouse. In the meantime, sixteen mallets were brought in to cover train operations until replacement parts were received and repairs made. The possibility of power loss was finally eliminated in the next few years when an additional generator was added to the power station and the system was tied into Virginian's electric lines at Matoaka. Further disruptions in service would thereafter be covered by Virginian's powerhouse at Narrows, VA.

Two other small incidents occurred later in the course of operations. An oil-type circuit breaker exploded on the second unit of LC-1 locomotive 2501 on July 20, 1927 at Eckman yard. One crewman was injured. Investigation determined a ground in the cable running between breaker and the transformer caused the accident. In the second accident, the vertical handhold on the cab door between two locomotive units in Bluefield yard gave out at the top end on April 30, 1931. This caused an employee to fall to the tracks between the moving units, but quick thinking prevented more serious injury as the worker laid within the gauge of the rails until the locomotives passed over. Investigation determined the bolt securing the top handhold had broken off at an old fracture point.

The onset of the Depression had a drastic effect on the coal business. As activity fell off, N&W reviewed railroad operations looking for economy measures. Expenditures for equipment were curtailed and work consolidated. N&W's operating ratio had declined with the start of electric service. Still, the company looked for a reduction across the board. The workforce at Bluestone underwent review. As an economy, the wheel shop at Bluestone Junction was closed in 1932. This work was consolidated with the shop in Bluefield where wheelwork on steam power was performed.

During the Great Depression. coal tonnage bound for industrial purposes fell off most sharply. Traffic declined to less than half its former level by 1932. The depressed economic conditions eliminated the revenues needed to fund expansion of electric train service. The existing surplus of steam motive power was now available to handle this work. The planned Williamson Electrification Extension was left uncompleted. In 1931, N&W extended a line from Wharncliffe to Gilbert. This reduced coal traffic han-

Although coal traffic fell off during the Great Depression, the economy of operation afforded by electric train service became more important to the railroad. The 2515, one of the more powerful LC-2 engines, is in charge of a train near Maybuery, WV on June 12, 1930.

dled by electric trains through the Virginian interchange on the Bluestone branch. N&W's operating ratio began to climb as a larger portion of tonnage moved under steam power. Further expenditures were eliminated which more than offset this increase.

Gradually, the economy recovered. By 1937 eastbound coal was again moving at pre-depression levels. On the eve of World War II, Pocahontas Division dispatchers in Bluefield were coordinating about 125 train movements a day. This dispatcher worked closely with the Bluestone Power Director who regulated the system. The Power Director had to determine the requirements for current so that sufficient electricity could be generated at the power station. He also had to know the number of trains on the line, and whether they were moving up or down grade as this affected electric power needs.

An incident occurred which dramatically illustrated the limits of the system. Once, when the company telephone system was out of order, a two-engine eastbound at Farm, another 2-unit motor and helper at Eckman, a westbound train at Wilcoe, and a two engine set at Bluestone all started at once. The resulting overload blew out every circuit breaker on the mountain, causing the railroad to be tied up until everything could be reset. To prevent a recurrence of these events, a special rule was written into Pocahontas Division timetable instructions reading: *"...All freight trains eastward, with two electric engines, will get permission from the Power Director before starting between Farm and Coaldale. If unable to get Power Director on account of the phone being out of order, it will be permissible to proceed without instructions".*

The job of dispatching the electric lines also involved consultation with yardmasters at Bluefield, Eckman, Wilcoe and Kimball in order to regulate movements. Sometimes the dispatcher would be talking to three yards at one time, while monitoring other activities. All of this called for experience and a "cool head." The dispatcher's assignment called for combining of trips to the coalfield to set out empties, with the need to move accumulated loads, while taking into account traffic passing through the division. This job was made easier by double track, replacement of tele-

Power Director Weaver is seated at the control board at the Bluestone power house. Information from the dispatcher, yardmasters, and train crews help determine power requirements for operation.

By 1937, coal tonnage began to rebound sharply from the effects of the Great Depression. The more powerful LC-2 locomotives held down a large portion of this business.

graph with telephone and especially automatic signaling.

Maintenance of the electric catenary system went on continuously and involved a diverse number of crafts. Two groups of linemen did line maintenance, one with its base at North Fork, and the other located at Bluestone Junction. Their duties included repair of damage, painting of structures and testing of insulators. A procedure was developed for employees to qualify to work in the maintenance and repair of the system. The highest were the general line foremen who were judged competent to erect, repair, and maintain electric apparatus, or to supervise and protect other persons working in the electrified territory. Below these employees in skill requirements were trainmen, engine men, car inspectors, and other approved employees in electric service. These were permitted to go on top of high equipment. The final groups of employees were supervisors, assistant supervisors, track foreman, and employees who were permitted to operate machinery in Maintenance of Way service.

By its very nature, electric operation required constant maintenance. Unforeseen at the start, and a major factor in its later elimination, had been the nearly continuous modification in the overhead catenary caused by the steady rise in track elevation. This averaged about 1 in. per year and was necessary because of the steady accumulation of a large amount of coal dust in the ballast that was dropped from cars. Forking the stones did not remove this, leaving the most economical solution being to reballast the track at a slightly higher level. After ten years of operation it had been necessary to raise the trolley wire several times to maintain proper clearance. In most instances this was accomplished by shortening the length of the hangers. The procedure continued until at some points of low clearance this remedy was no longer possible.

Maintenance forces then moved to the solution of lengthening the poles. This was an expensive and labor intensive procedure. It was accomplished by first cutting off the tubular poles at the ground level with a torch, then resetting them in a casting that consisted of

LC-2 2513 rolls eastward under the catenary and signal near Bluestone, WV in 1930. The signal system was essential to operations and had been in place since the start-up of electric operations.

a base plate with a cylindrical sleeve. A cement grout was used to fill in any distance between the pole and the sleeve. The base plate was attached to the concrete foundations with two anchor bolts, which were long enough to permit the base plate to be extended 8 in. The extended ends of the anchor bolts were then covered by pipe caps and held in place by extra nuts.

Upkeep of the catenary and associated equipment was necessary for safety and efficient operation. The chief sources of trouble were wrecked pantographs and insulator flashover. Several general line foremen were trained and placed in charge of all facets of repair and maintenance of the catenary. They knew all employees assigned to work on the wire train and were responsible for the strict observance of special and safety rules under which work took place. In accordance with these procedures, the general line foreman obtained necessary clearances for all work to be performed or assigned another employee of equal competency to do so. The general line foreman for a specific repair or other project determined which circuits would be de-energized, the extent of clearance, the locations, and the nature of the work to be performed. Should the general line foreman have to leave the immediate work area, he would assign another employee of equal ability to take charge until his return. He would also advise each man in the gang the name of the person assigned to take charge as a safety precaution.

The gasoline powered motor car was the most commonly used transportation in maintenance work. These had sufficient power to pull a trailer carrying materials. Work was also done from these cars, and the wire train was only called out once a year, excepting emergencies and wrecks. A wire train was maintained at Bluestone Junction, with specially equipped cars to facilitate this work. The wire train was a self-contained work unit, which could be moved to any site needed. It was kept in a constant state of readiness in order to respond to any accident, and performed special maintenance when not otherwise engaged. The most visible piece of equipment was the high wire car. This car was equipped with a grounded pantograph and featured a high platform from which inspection and repair work to the trolley took place.

Work was done on hot or active wires only in the spans in order to guard against accidental grounding. Special 12-ft. ladders were impregnated with paraffin and used in hot wirework. When work was to be done in the insulators, it was necessary to have the line deactivated. The general line foreman had duties after he obtained necessary clearances. He explained to all men involved the designations of the de-energized circuits and the extent of the clearance obtained. Each man would repeat the information received. The general line foreman would personally direct the raising of the grounded pantograph, a safety device, and observe it strike the contact wire. He would be the first person to ascend to the top of the equipment.

Once on top, he would direct the application of the required grounding devices. He would direct the attention of each man to the location of energized circuits in proximity to the work. Each man was required to repeat the information received. In so far as practicable, the general line foreman would locate himself so that he could observe the movements of all men on top of the equipment at all time. Work could then proceed to inspect each insulator for wear and cracks. Each insulator was tested with a "megger", which detected a current leak, every one or two years. The replacement rate was 2% per year.

Upon completion of the work, the general line foreman would direct the removal of grounding devices and be the last man to descend from the top of the equipment. He would immediately advise all men that the circuits were about to be energized. All employees, regardless of their assignment or duty, had to report to the foreman before going on top of the equipment. When adjacent wires were energized and it was necessary to work in close proximity to the energized circuits, the general line foreman was personally required to supervise the work. He would also prevent tools, apparatus, or employees from approaching within 14 inches of the energized wires. Under no circumstances would any employee other than one judged competent to erect and repair electrical apparatus be permitted to approach within 3 feet of wires energized at 11,000 volts. Few accidents happened in line work.

Another source of work was the wooden strain insulators that were used in the pull-offs and between tracks. Initially these gave occasional trouble from flashover but this was stopped by re-insulation. These wooden insulators were recycled after having the original varnish scraped off. The insulator was then heated in a baking oven to dry out. It was then repainted with a dipping varnish and baked, after which it was reused in service. A year after reinstallation, wooden pull-offs were given another coat of air-drying varnish.

Employees assigned to maintain the signal system received less training in areas of electrical theory and repair. Still, this occupation was essential to train operations. The semaphore signal system had been in place since startup of electric operations in 1915. At that time, a relatively new D.C. powered system was taken down so that a newer system that would conform more readily with A.C. powered train operation could be installed. The 44,000-volt power line for train operation was carried along on the south side of the catenary tower, while two 4,400 volt signal power lines were carried on the north side. The two lines, duplicates of each other were each capable of carrying the entire signal load. All signal control and associated telephone and telegraph lines were also carried at the northern edge of the property and maintained by the signal department.

In order to protect against impedance interference from the 44,000 volt line, the signal transformers were located closer together than would otherwise normally be installed. These were checked during routine patrols. The signal transmission line was carried on porcelain insulators and broken at two-mile intervals on strain insulators. At these points, power was carried through a double-pole, single throw, oil insulated switch. Workers could open each signal line section for testing or repairs. In addition to the signals, power was provided off the 4,400 volt feed for lights at stations, towers and yards. Power was also taken for a motor driven air compressor at Elkhorn tunnel, and a low wire warning light, (for locations where the trolley wire was low).

A 1 k.v.a. 4,400-110 volt line transformer, with 5% and 10% taps for line-drop compensation, was used at each signal location. At the very few points where more current was required, a 3.5 k.v.a. transformer was installed. Lightning arrestors and fused line cut outs protected each transformer. These signal transformers were mounted on the same poles that carried the 4,400 volt feeder line. Screens were placed under the power wires, so that persons on the bridges could not accidentally come in contact with them. A hand railing extended all the way around the signal platform on the bridge and a ladder provided access from the north side, extending down to within 7-ft of the ground. It stopped here to deter unauthorized access. Signal repairmen were issued a short extension ladder as part of their regular equipment, which connected to the catenary tower ladder.

At each westbound signal location, the common wire was broken and small one-to-one transformers were placed in the control circuit. These minimized the effects of electrical interference, mainly induction, and were inspected during routine motor car patrol. Switches were installed so that current could be cut off while the signal maintainer was at work.

Double track circuits were used throughout the electric territory, the average block being about 4,500 ft.; although one particular circuit in an interlocking plant was as short as 300 ft. All insulated joints were six hole Keystone type joints, with Bakelite canvas insulation. The track transformer, track relay, reactance coil and line relays were housed in cast iron relay boxes on concrete foundations at the base of the catenary bridge where the signal was located. Track wires were carried from the relay boxes to the tracks in fiber conduit, and to the signal transformer and signal line in galvanized steel conduit.

Seventy automatic signals were required for the initial Elkhorn electrification, with more added for the extension. They were United Switch & Signal Company's S-style upper quadrant semaphore operated by 100-volt 60-cycle induction motors. The signals were mounted on top of the bridges with special attention paid to location, in order to bring about the elimination of obstructions. The signals were enameled and equipped with Armspear spheroid lens lamps due to the excessive curvature of the line. These signals, along with interlocking towers located at strategic points, the power director, and yard masters served to keep the electric trains in motion.

Typical of the type of interlocking towers in the electric region was the installation at Graham, VA, located just outside of Bluefield. The tower controlled entrance to Bluefield yard for both the electrified route to the Pocahontas coal field and the steam powered Clinch Valley line. The interlocking plant had 14 working and 2 spare mechanical levers, as well as six electrically powered levers. All control circuits were D.C., operated by a battery of five lead cells. These were in turn charged from a mercury-arc rectifier taking power from the signal line. Approach and route locking were provided for all through movements at the interlocking.

Graham Tower, located at the junction of the Clinch Valley District, has 14 working and 2 spare mechanical levers, in addition to 6 electrically powered levers.

A special problem for mechanical forces was the difficulty in clearing train derailments in electrical territory. Because of the danger of accidental electrocution special rules existed. Should a derailment occur, an employee judged competent to erect or repair catenary was dispatched to the site, where he reported to the trainmaster and wreck foreman. Should the wreck be of a serious nature, a wire train was dispatched so that major electrical work made necessary by the wreck could be performed. If the catenary was to be de-activated the Power Director was contacted along with the dispatcher. The dispatcher would then make available steam locomotives to retrieve cars from or power the wire train near the immediate site of the wreck.

The Bluefield wreck derrick was used in electric territory. Its boom was changed out with another derrick, and a shorter boom substituted in order that it could more easily work in confined quarters underneath the wire. It was also equipped for at least part of its life with a wire guard on the boom. This was used to protect the wire when it became necessary for the wreck derrick boom to come into contact with de-energized or grounded wires. This sometimes resulted in upward or lateral pressure that could damage the tension of the wire. Movement of the boom could only proceed under the direction of a general line foreman.

The wreck derrick could be operated within 8 feet of the high tension transmission line and to within 3 feet of the catenary or signal power lines without a general line foreman. With this employee present, the wreck boom could be brought closer, but only if the wires were de-energized and properly grounded. With the catenary electrically isolated at wreck sites, confining area under the wire still hindered repair work. Wrecking cranes could not fully extend their booms to lift locomotives and cars, making this an inherently unsafe assignment. The catenary could not be re-energized until the qualified employee in charge made sure all employees were clear, at which time the Power Director was contacted.

Overall maintenance of the catenary infrastructure, substations, power stations and equipment was a constant process. Although motive power work was not as labor intensive as steam, right-of-way work under the catenary was more involved. The crafts all worked together as an integral part of electric operations, each necessary to keep the coal traffic moving. Still, with age, maintenance costs on the electrification began to rise, a fact that did not go unnoticed in Roanoke.

During the Second World War the electric locomotive fleet kept the coal traffic moving. Tonnage at first dropped on account of German submarine activity off the Atlantic seaboard, then rebounded as American industrial productivity increased with the coming conflict. The LC-2 motors were now handling nearly all road assignments. The LC-1 engines, now nearing 30 years of age, saw service in helper duties, and crept into head-end train duties when LC-2s were otherwise engaged.

Traffic on the Pocahontas Division rebounded with the passage of war-bound material for Norfolk from midwest points. Norfolk, site of a major U.S. naval base, was also a port of embarkation for men and equipment heading for the war zone. Added to this crush of war material was the movement of increased coal traffic necessary for stepped up industrial production. Norfolk & Western was also obligated to maintain some semblance of normal freight and passenger schedules throughout this period.

Coal traffic continued to increase throughout the war. In 1943, the railroad moved a record 54 million tons of coal with the electrics working throughout the conflict to keep the traffic moving. The unprecedented business levels took its toll on the motive power. Keeping both the LC-1 and LC-2 class engines in service proved costly, with the motors requiring more frequent and expensive repairs.

Following the conflict, changes took place in the upper reaches of Elkhorn Creek that had a lasting impact on train operations. Years of mining activity had removed the easily recoverable coal, leading a number of companies to reduce or close facilities. These shifting mine operations opened possibilities for N&W to again consider making line improvements. Relocation studies were undertaken with the idea of removing the heavy grades and sharp curves that handicapped the Elkhorn grade. These studies intended to lower costs of operation, and were to be accomplished by taking advantage of new low-cost advances in construction techniques and available space reclaimed from closed mining sites on which N&W could now relocate the mainline . The grade and curvature problems on the west slope were the most pressing concern for the railroad. Concurrent with these pending changes, some decision would be necessary concerning the continued operation of electric train service.

The railroad was fast approaching the mid-century mark. N&W was still relying on a motive power system put together before the first World War for moving coal. This system operated over a line constructed before the century as a coal feeder branch. Electrification had reached the end of a natural service life with both the locomotives and power stations now worn out. In 1915, the railroad had been able to double the overall thermal efficiency of the Elkhorn operation to nearly eleven percent by switching from steam to electric. Advantages of electric service at that time were an increase in tonnage and doubling of train speed. These values remained at fixed levels. In addition, line-side catenary maintenance was found more labor intensive than steam roadway work. N&W looked for some modern method by which further improvements could be made.

Electric locomotives had never replaced steam entirely on the Elkhorn. While eastbound coal and returning empties moved behind electric locomotives, westbound coal had always moved behind steam. All passenger traffic on the Elkhorn had remained a steam operation, with the exception of the occasional helper on heavy trains. Similarly, time-freight service moved behind steam. These operations had given the railroad continued experience with the workings of steam motive power over the grade. The company in the meantime had continued to refine and perfect steam, creating a new "modern coal burning locomotive." In 1945, this was epitomized by the Class Y6b 2-8-8-2, a vastly superior machine to the Class Z 2-6-6-2 that had lost the job to electric motors. The Class Y6b had a theoretical thermal efficiency approaching seven percent.

The electrification project used the best equipment and technology that could be assembled when first constructed. The state-of-the-art watertube boilers at Bluestone Junction, operating at 200 lbs. pressure, used between two and three pounds of coal to produce one kilowatt hour. This had declined over the years to where the power house boilers now required nearly four pounds of coal to produce one kilowatt hour. The 200-lbs. operating steam pressure was outdated and exceeded by nearly every steam locomotive on the post-WWII roster. Overall thermal efficiency of electric operation had dropped from eleven to less than eight percent.

The railroad decided to proceed with roadway relocation. The plan accepted called for the elimination of electric train operations and its replacement with steam. Studies had shown coal tonnage on each train could again be doubled, this time by using the Class Y6 steam locomotive on a new, relocated right of way between Lick Branch and Cooper. The study recommended boring a new double track tunnel at a lower elevation and relocating the roadway to decrease both curvature smd grade.

The cost of the improvement would be recovered by (1) the increased efficiency of the new steam locomotives and (2) curtailment of operating costs associated with maintaining or upgrading electric service. The plan for roadway improvements called for the project to be broken into two parts. Phase 1 involved roadway work on the west slope of the tunnel, while Phase 2 covered work on the east slope. Work on the west slope began first and continued after Phase 2 began to the east. Connecting the two projects would be a new 7,135 foot long double tracked Elkhorn tunnel.

Initial work was begun on Phase 1 on June 5, 1947 under the direction of Resident Engineer C .W. Fiery. By Christmas of 1948, Phase 1 on the western portion of the tunnel was about 99% complete. This work involved the relocation of 5.27-miles of line between Cooper WV, on the east, and Lick Branch on the west side of Elkhorn tunnel. In order to reduce curvature and lower the grade from 2 to 1.4% (compensated) between Coaldale and Maybeury, the line was relocated to the south side of the creek calling for a large fill leading to the location of a new tunnel. Curvature was reduced from 13 degrees to 4 degrees. Construction on the west slope also involved the building of a new 538 foot long bridge between Maybeury and Lick Branch, and a 200 foot long concrete culvert at Barlow Hollow to be covered by the fill approaching the new tunnel.

Work on Phase 2 of the project began on December 15, 1947 under the supervision of Resident Engineer Roscoe Porter. By Christmas of 1948, work on this side of the mountain was about 65% finished. Curvature was eased, but not nearly to the same extent as on the west side. The 1.2% grade on the east side was practically eliminated, except for a short stretch between Cooper tunnel and Bluestone Junction. The largest portion of this project was the rebuilding of bridge 854 at the west end of Cooper Tunnel, the new steel viaduct to be 690 feet long. Both projects together moved more than 4 million cubic yards of earth and rock. Linking roadway improvements on each side of the mountain was the new Elkhorn tunnel. The new double track tunnel was more than twice as long as the old bore, having an inside area more than 3 times as large. The grade against eastbound traffic inside the bore was reduced to 1% as compared to 1.4% in the old tunnel at Coledale. The firm of Haley, Chisholm & Morris of Charlottesville, VA performed actual tunnel construction. Representing the construction company on site was Mr. Robert Parker, whose previous projects included excavation for Grand Central Terminal in New York.

Work on the tunnel began with a geological survey over the site. This was necessary for a number of reasons, but most importantly to gather information on the character of the rock through which the tunnel would penetrate. In addition, there had been reported mining activity in the immediate area of unknown extent and nature. At intervals of about 1,000 feet, a series of test holes were drilled from the top of the mountain to a depth of 20 feet below the grade of the tunnel. Cores from these drillings were retained and records made so that engineers would know in advance what to expect as they worked the tunnel into the mountain. The shallowest of the core drills was 170 feet, progressing to 440 feet at the top of the mountain.

A special machine used for a drilling device was "Jumbo," which had been invented and assembled by contractor Haley, Chisholm & Morris. The machine was 36 feet wide and 35 feet high and speeded the drilling operation by allowing 35 men (drillers, helpers and foremen) to work on the tunnel face at one time. "Jumbo" operated on its own specially constructed set of wide gauge railroad tracks.

Holes were drilled by "Jumbo" into the rock facing to which blasting charges were loaded. The machine was then pulled back on its track and the face of the rock was shot down. Next, shovels and trucks moved in under "Jumbo" and brought out the material, known as muck. "Jumbo" then went back into position and placed steel liners into the newly cleared area for forms. Concrete was poured into these forms to line the tunnel, and the new facing was again drilled for blasting charges as the process was repeated all over. By this procedure, work using 15 drills on "Jumbo" proceeded at first from the west end of the tunnel. Later, a second machine was set up at the east end, both proceeding at the rate of about ten feet per day. A total of about 315,000 cubic yards of rock were moved from the tunnel.

A second drilling machine, working from the east face deposited soil along the new right of way in the hollow downgrade from the portal. The old line to Elkhorn tunnel can be seen on the hillside to the right.

Roger F. Whitt Collection

The eastbound mainline has been laid in the new Elkhorn Tunnel and the ties put in place for the westbound track on August 22, 1950.

The tunnel was lined with 70,000 cubic yards of concrete and reinforced with 1.1 million pounds of steel rod. The sides and roof of the tunnel were reinforced with nearly 14 million pounds of steel liners. 26,000 feet of drain tile were installed to carry off moisture from the tunnel walls and 85 thousand feet of conduit were placed to carry signal, telephone, telegraph and high tension lines through the bore. By the spring of 1950, the tunnel was holed through as the two drilling crews met under the east slope. As concrete work had proceeded concurrent with drilling, little additional work was necessary before track work began. The eastbound track was laid first with track crews using the westbound track space for M-of-W equipment. After completion, the westbound track was installed.

The railroad planned to open the new tunnel prior to completion of the entire roadway project and organized a celebration accordingly. A plan was formulated to have the company's most prestigious passenger train, the *Powhatan Arrow,* inaugurate service through the bore. The opening was a big event for citizens of the whole area. On June 26, 1950 a crowd of about 100 people met the *Arrow* at Welch to give the train a big send off. Flowers were presented to engineer N.G. Danner at the station stop by Roy B. Shrout, representing the McDowell County Chamber of Commerce. Then, about 50 citizens, including Mayor-Elect B.F. Howard boarded the train to journey eastward to the tunnel.

Similarly, a crowd of about 300 onlookers, including representatives from the railway, the contractors who built the tunnel, and newspaper and radio personnel gathered at the east end near Cooper, WV. At 3:48 P.M., the eastbound *Powhatan Arrow* officially opened the new Elkhorn tunnel. A red ribbon emblazoned with silver N&W monograms over each track was stretched across the mouth of the tunnel. As Class

Service through the new Elkhorn Tunnel was inaugurated by The Powhatan Arrow, *the railroad's premiere passenger train.*

Construction proceeds over the new bridge over the Bluestone River at Cooper, WV while an electric helper passes westbound on the old line. Below, the remnants of the original line and "Guard Rail" curve can be seen below at the left of the photo.
Roger F. Whitt Collection

J locomotive 603 broke the ribbon, a cheer went up from the crowd.

On board the train, the regular crew of engineer Danner, fireman Ernest T. Epling, conductor J. R. Doyle and trainman R.H. Lindsay was joined by Train Passenger Representative Pillis. Pillis gave a running account of the project as they passed through the area east of Lick Branch. Passengers scurried from one car to another to view the proceedings. From their windows they saw people standing all along the new line to watch the opening. Large crowds were present at both ends.

A portion of the opening ceremony was carried on Bluefield radio station WHIS, which interviewed officials from both N&W and Haley, Chisholm and Morris. Although the tunnel itself was complete, only eastbound traffic initially operated through the bore. Much grading, track and signal work remained before the westbound track could be opened. By completing the eastbound line first, coal trains could take immediate vantage of the easy grades. Westbound traffic used steam and the old single track Elkhorn tunnel located 3,000 feet to the north.

Completion of the new bridge 854 over the Bluestone River at Cooper marked both the end of electric train service and the opening of the new westbound main. The new double track line continued as far west as the Maybeury bridge, where it switched back over to the old line, pending roadwork further west. Rail was taken up through the old Elkhorn tunnel and the facility turned over to mining. Remaining seams of coal were robbed from the walls of the tunnel before the work was finally abandoned.

On the same day that the new Elkhorn tunnel was placed into operation, June 26, 1950, the catenary was deactivated. N&W's electric fleet languished at Bluestone Junction after the electrification was decommissioned. Power for accompanying signals, lighting and telephone lines was cut over from its source at the Bluestone power plant and fires in the boilers killed for

The east end of the new Elkhorn tunnel shortly after completion of the project. The old tunnel was located to the north (right), and about 100 feet higher.

the last time. N&W's Class LC-1 electric motors were completely worn. The newer LC-2s, which had performed admirably during the conflict, were not in much better shape. When no buyer stepped forward with a new home, the electric locomotives were scrapped.

The railroad's quest to find a solution for moving coal over the Elkhorn grade had taken it full circle, going from steam motive power to electric, and back to steam again. Within a few years, little trace could be found that electric train service on the Elkhorn ever existed. Electric traction had successfully met one of railroad's toughest challenges in mastering N&W's coalfield mountain crossing. ■

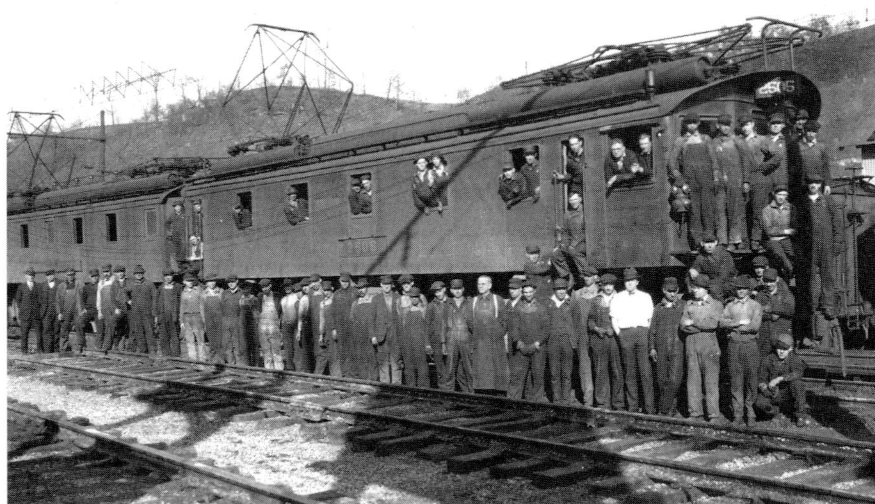

Employees stand for a photograph with an electric lomotive at Bluestone, WV.

6 Indian Summer

Norfolk & Western inherited a first-class electric operation when it merged with neighboring Virginian Railway. On April 4, 1961, only a short time after the merger, former Virginian EL-C locomotives prepare to depart from Elmore with a train for Roanoke.

On December 1, 1959, Norfolk & Western inherited a thoroughly modern electric operation through merger with the neighboring Virginian Railway. By that time, the railroad had been out of electric operations since 1950, having replaced its Elkhorn grade electric operations with both line relocations and steam motive power. Since then, further changes on the Norfolk & Western found diesel-electric motive power replacing steam in train operations. But once again, N&W found itself back in the electric train business, as it brought the partially electrified, partially dieselized Virginian Railway into the family.

Henry Huttleston Rogers had constructed the Virginian in the early part of the century, almost expressly for the purpose of moving coal out of the West Virginia hills. The Virginian Railway was engineered and constructed to move heavy tonnage at the lowest possible operating expense. The line was built from Deepwater, WV through the New River and Pocahontas coal fields eastward to Norfolk on the Atlantic seaboard. Despite the high engineering standards brought to bear, the railroad traveled through a particularly difficult terrain. This area had a complex topography, that consisted of a series of lateral mountain spurs with steep valleys. The coal deposits were located in these narrow valleys. The west end of the Virginian Railway on which this coal originated featured a difficult operating profile, with many bridges, tunnels, curves and grades. Little room existed for additional railroad sidings or running tracks.

Eastward from the main coal collection yard at Elmore, WV, the railroad grade exceeded 2 percent, with curvature approaching 12 degrees on the climb out of the coal field at Clark's Gap. Double track was applied early on this hill in order to ease congestion on the ruling eastbound grade for the entire railroad. In addition, the track was ballasted with rock and laid with rail at 130 lbs. per yard. It featured steel bridges that were built to the highest loading standards.

The Virginian remained at the forefront of steam locomotive development. It received its first 2-6-6-0 in 1909, shortly after the mallet design was introduced in the U.S. Still larger engines joined the roster; a series of 2-8-8-2s, massive 342 ton 2-10-10-2s, and an experimental 2-8-8-8-4 in 1917. The difficult 2.11% grade to Clark's Gap determined new motive power purchases. The railroad routinely moved some very heavy tonnage trains over this grade. Normally operating with a Class US 2-8-8-2 on the head end as a road engine, and with two Class AE 2-10-10-2s pushing, 5500 tons could be taken up the grade to Clark's Gap where the pushers cut off. The typical train filled out to 8000 tons on the east side of the hill at Princeton, WV, where the road engine got an assist from another pusher out of the yard. A pusher was added again further east on the 0.6% grade up to the summit from Whitethorne to Merrimac, VA. Moving coal trains of this tonnage over similar grades was unheard of elsewhere, but was an established practice on the Virginian.

At the end of World War I, the coalfield served by the Virginian began to mature and became profitable. By 1920, the carrier was moving approximately 7 million tons annually, an amount expected to double in the next few years. A comprehensive plan for improvements was undertaken. The necessity of having to add to the locomotive fleet was soon realized, with large expenditures for facilities to house, coal and water the engines. But perhaps the most costly aspect was the need to double track its entire mountain line. The railroad studied all improvement options, including electric traction.

Looking at the success of N&W's electric service, it seemed reasonable for Virginian to anticipate that extensive double tracking could be postponed if electric train operations were introduced. Electric operation on the N&W had speeded up the entire railroad, creating additional track capacity. To equip the Virginian for electric operations would cost about 15 million dollars, a very large sum of money. The net investment, when taking into account the necessary new expenditures for added steam capacity, fell to 5 million dollars. In addition, the expense of further double tracking would be unnecessary. The real savings to be produced by electric haulage over steam occurred in fuel, time of train crews and better utilization of equipment coming from increased train speeds, hauling capacity and reduction in delays. It was finally determined that electrification would be profitable due to these savings, and offered more flexibility and reliability as well.

Virginian retained the engineering firm of Gibbs & Hill, with both Mr. George Gibbs and Mr. E. R. Hill personally directing the design and construction work. Engineering work commenced on June 1, 1923. Construction of the first part of the program to electrify its mainline from Mullens, WV to Roanoke began on November 29, 1923. The electrification covered a distance of 133.6 miles. As usual, Virginian did things in a big way. Gibbs & Hill's assignment involved adapting a heavy electric system to Virginian's transport system. This could only be accomplished by making a number of changes in existing steam railroad facilities. Certain operating methods needed to be modified to obtain maximum benefit from electric operation. To this end, Gibbs & Hill made much use of Virginian operating records, right of way plans, and grade plats to calculate these values. Conferences were held with company officials to determine further electric train use and what additional facilities would be needed.

Design work used Virginian roadway cross-sections, track charts, drawings of tunnels, viaducts, buildings and line side structures which might affect the clearance of catenary structures. Each was checked in detail with some supplementary survey work done as designing progressed. Data was gathered on sub-soil conditions in order to determine the depth and character of catenary column and guy-anchor foundations. Survey work was required since the high voltage transmission line did not follow the right of way east of Clark's Gap. Instead, it had been determined to be more economical to run through the mountainous and heavily wooded country on a straighter course, as Virginian had indicated at some point in the future it may relocate parts of its right of way. Negotiations with landowners for purchase of property or easements were conducted.

To facilitate the work, Gibbs & Hill established a field organization with Mr. H. W. Hudson placed in charge as Resident Engineer. He directed operations from a centrally located headquarters on the line. Besides his employer, Mr. Hudson reported to several Virginian Railway officials including H. Fernstrom, Chief Engineer; Hugh Patterson, Engineer of Electric Traction; and John W. Sasser, Superintendent of

Motive Power. Mr. Hudson had charge of all surveying parties. Later, when construction started, he organized and carried out instructions from the home office in New York. Many special circumstances were encountered, especially in the design of the electrical distribution system for 137 miles of track. Almost every catenary span had to be dealt with as a separate problem. Variables included local ground situations such as the foundations, track curvature, clearances and number of tracks.

Mr. H. S. Richmond served as Structural Engineer and headed up the work assisted by a team of designers in producing more than 3,000 drawings necessary for the entire installation. A time-saving aspect of the designs was the reduction of the amount of field cutting and fitting of pieces for the catenary system during the process of erection. Information gathered from the railroad concerning track alignment and elevations was used in laying out special drawings, particularly on curves and distinctive overhead work such as track turnouts and electrical section breaks. Information was prepared for the field forces for each particular catenary span which included cable sags and hanger lengths. This work was done prior to the distribution of materials, resulting in a saving of time.

The system selected was not identical to Norfolk & Western's electrification. Although the same 11,000 volt, 25-cycle A.C. was supplied to the trolley through a steel H-section catenary system, it was more sophisticated than N&W's through the use of a "balancing" wire and transformer system. The Virginian catenary was designed to deliver the large amounts of power, handle higher voltages necessary for heavy train operation on the steeper grades, and offer a greater degree of protection in the event of short circuits and power surges.

Gibbs & Hill had established construction headquarters at Narrows, VA, where the power plant was being built. Secondary headquarters for catenary construction was established at several other points along the right of way. These were generally located at camp train sites, which became mobile living quarters for the men. These camp trains could be easily moved to the point of construction. Five of these trains were used at certain times during the work, and each train differed according to the work assigned. Tent camps were provided for the surveying and erection forces assigned to the high-tension transmission line, since this work was remote from the railway. Tower cars were used in the train assigned to the erection of catenary wires and fittings. These tower cars were built in the Virginian shops from N&W plans. Wooden frames were erected on these cars to serve as high level working platforms on standard flat cars. A number of automobile-type gasoline motor cars were used to haul air compressors for field drilling. The adjusting and tensioning gangs later used them for inspection and movement of men and light equipment. Their occupants could quickly remove these lightweight motor cars from the tracks when necessary to clear for a passing train. Some work was let to local contractors, including the grading required for electric sub-stations.

All construction was carried out so that it did not interfere with the movement of train traffic. The high voltage transmission line consisted of assembled steel grillwork with the footers set in earth excavations and then back filled. This line did not follow the right-of-way and was laid out to avoid excessive handling of the heavy pieces in difficult country. The erection of tower sections was done with the use of a gin-pole carrying a block and tackle. Horse teams were used in this part of the work for haulage and hoisting power. The wire

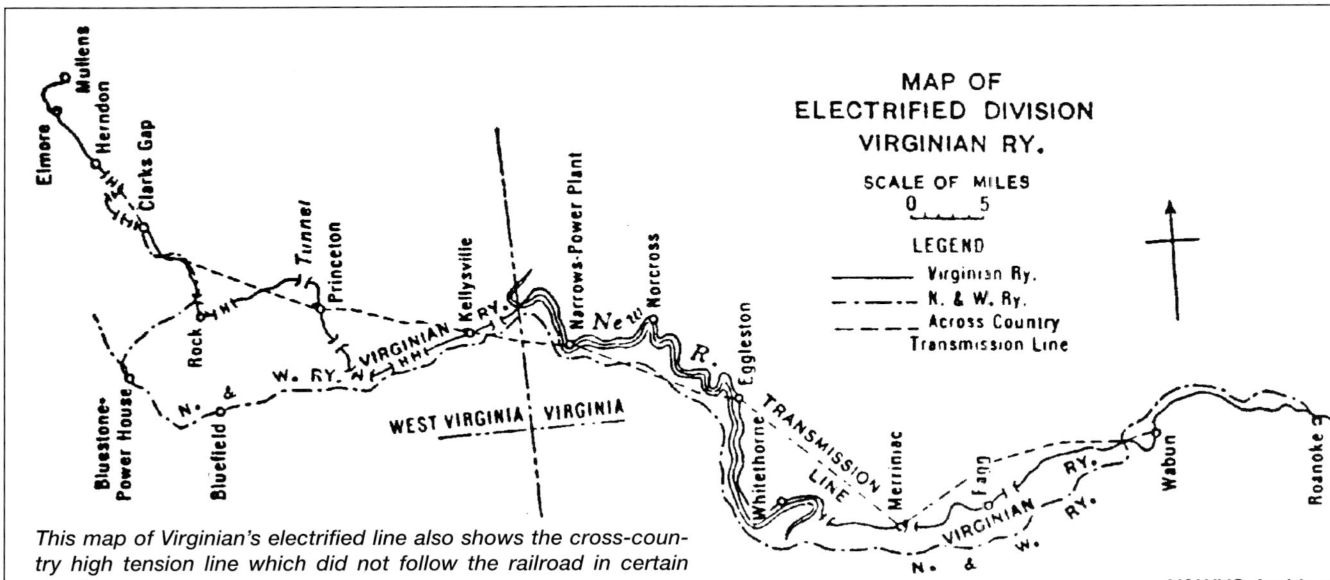

This map of Virginian's electrified line also shows the cross-country high tension line which did not follow the railroad in certain locations.

N&WHS Archives

cable for the high voltage circuits arrived on reels, each containing about 2600 feet. They were hung through snatch blocks attached to the tower cross arms. The cable was sagged using block and tackle. Proper sag was determined by sighting through targets on the towers and on cables at mid-span. About one tower was erected and 6 miles of cable strung and sagged daily.

Catenary construction was done taking into account regular traffic patterns. Virginian catenary construction used a method which eliminated the concrete work trains. Instead, the foundation holes were dug and blasted out, then fitted with corrugated steel casings. The casings were back filled and capped, then left ready for quick placing of poles. The steel H-beam poles were placed into position by a locomotive crane assigned to a following work train. By using this method, as many as 35 poles could be set in a 10-hour working day. The guy anchors were set in the same way. When the poles were placed in the foundations, wooden blocks were used to wedge them plum. The wedges were removed when the poles were backfilled. After the pole was set, the structure was secure. Use of a work train and crane was unnecessary to finish the erection.

Wires on the catenary system were strung in the following order by use of the extension-top tower cars operating in another work train: first, the balancing transmission wire was installed, then the steel cross-spans, the messenger wire, and finally the trolley and auxiliary wires. Proper tension was applied in accordance with design specifications by using a dynamometer. An average of four reels of messenger cable, each with 4500 ft., were strung and sagged in one day, but varied depending on traffic conditions.

A separate gang distributed and erected the hangers, which had been pre-cut as determined by the design team. The hangers were pre-fitted with top and bottom fittings. The degree of accuracy was such that hardly any field cutting was required. Hangers were applied to the messenger cable on top by a construction worker riding the messenger wire in a "boatswain's chair". Following behind and working from the tower car, insulator strings were installed by other construction workers, the auxiliary and trolley wires were clipped in and final adjustments made. Work normally went quickly with the work train clearing as needed for revenue traffic. At certain points such as overhead switches, section breaks, deflectors, and pull-offs, the operations were complicated and progress was slower.

Virginian decided to build its own powerhouse to supply the railway electrification. With the nearby availability of high-grade coal, a steam plant was determined to be the most suitable and economical. As an alternative, the railroad did consider purchasing power from commercial sources. This would have saved the railroad the projected 4 million-dollar cost of a generating station. Four different proposals were examined, but only one company submitted a definite proposition. This option was rejected as cost and terms were attached that made it impractical. Expected load conditions for Virginian were so exceptional that any local supplier who also served the general public could not afford to take on the railroad as a customer. It would

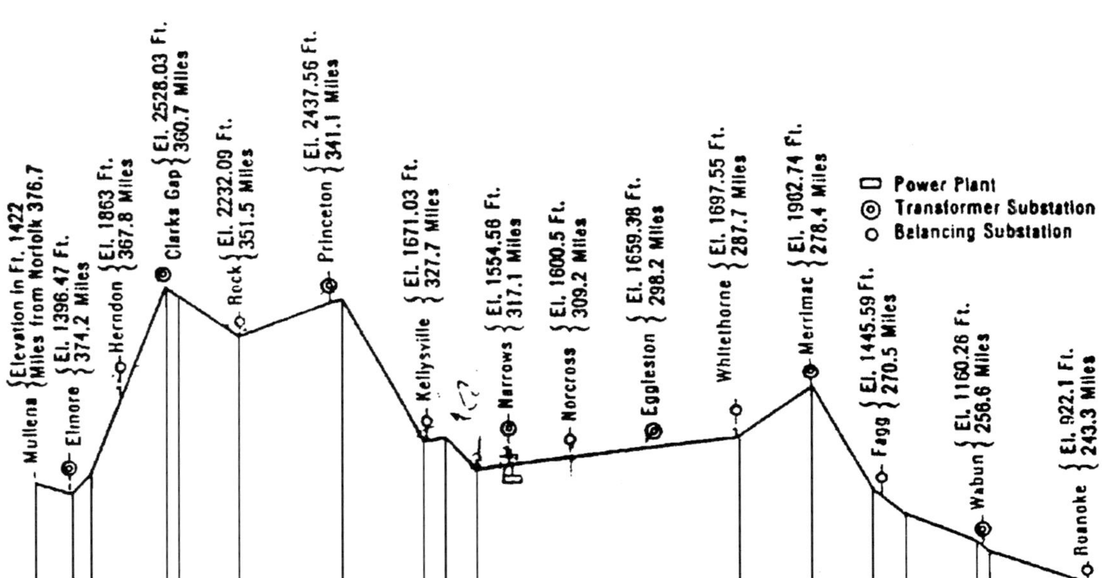

Distance Miles	2.478	2.38	10.28	0.967	8.535	13.6	1.034	2.535	4.15	33.801	9.44	7.225	4.24	8.9	1.602	13.02	1.0315
Max. Grade Per Cent	0.329	1.21	2.11	1.25	1.24	06.0	1.50	0.201	0.808	0.20	0.60	1.50	0.90	0.90	0.90	0.60	0.17
Max. Curvature Degrees	12	12	12	6.5	12	8	8.16	2.5	3	9.25	8	8	4	5.18	6	8	7.5

This track profile shows the location of the powerhouse and substations on Virginian's electrified lines.

N&WHS Archives

have only been possible by devoting special generators and transmission circuits to the task. The charges for this would have been passed directly to the Virginian in the form of increased rates.

A site at Narrows, VA was picked, being located midway between Mullens, WV and Roanoke, VA. The site was also near an established town where employees could be housed. Narrows VA was adjacent to a ready supply of water from the New River, which was necessary for boiler-feed and condensing purposes. The New River varied in volume of flow, but contained ample water for the power station at all times. Maximum floods occurred in 1870 and in 1916 with the site selected for the power station being above these flood plains.

Local rock under the site was limestone of variable quality, embedded to be subject to underground water erosion. The site was examined by test boring to determine sub-surface conditions. This was necessary as concentrated column-loading for certain sections of the power station were heavy. Considerable variation in the quality of the limestone was found and unfortunately for Virginian, unstable portions occurred at several heavily loaded points. No caverns were found below grade but the company had to establish a low-unit pressure of 3 tons per square foot for the smokestack, which was the most heavily loaded portion of the plant.

Special considerations had to be made in the design of the powerhouse, placing the basement 30 feet below high water level. This was done in order to keep a reasonable height for the main floor and to permit ready access for equipment and to supply water economically. This low basement level necessitated the construction of very substantial walls and floors under the turbine room in order to withstand the high hydrostatic pressure that occurred during flood stages. It had been calculated that at the basement level, the hydrostatic pressure produced an uplift greater than the total weight of the construction above. It was necessary to alter the design to insure against movement of the station in this event. This was accomplished by installing a reinforced basement floor having heavy concrete slabs between the piers of the turbine foundations. These piers were anchored with 134 1-1/2-inch steel rods grouted into the surrounding rock. To further aid against uplift, vents were provided in the concrete floor. The basement walls were built of reinforced concrete with the building walls constructed of brick and designed to be self-supporting. The floors, overhead

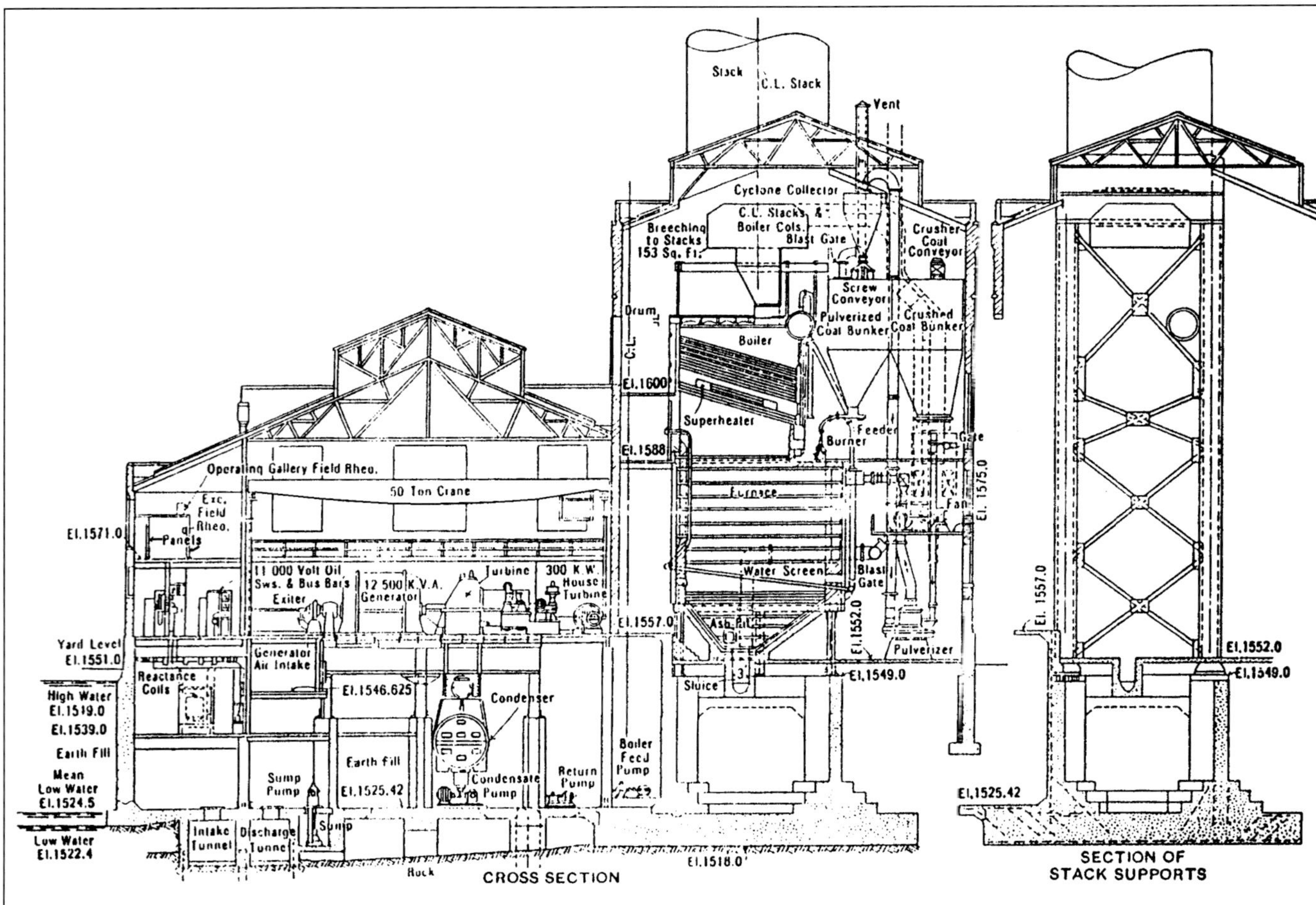

The general layout of the Virginian powerplant at Narrows, VA. *N&WHS Archives*

crane, stacks and boiler-room apparatus inside were carried on an interior steel framework.

The powerhouse was equipped with two stacks, each of which was constructed with reinforced concrete, supported on and directly bonded into concrete mats at the boiler room roof level. Four steel columns supported each stack. The columns served double duty as boiler supports. Flexible members tied these to adjacent boiler columns. In this way, the supports (and stacks) were permitted to sway without damaging the integrity of the roof. The main stack foundations rested on concrete piers, which in turn were on spread concrete footings. By this method of spreading the weight, the ground-loading limit of 3 tons per sq. ft. was maintained. The stacks rose 376 ft. above the bottom of the ash pits, and 285 ft. above the roof mats. The stacks were 25 ft., 8 in. in outside diameter at the bottom, and 16 ft. in diameter at the top. The design was able to handle high boiler ratings with natural air draft only.

A thirty-five car siding was installed by Virginian for coal delivery on the west side of the plant. Coal was delivered into a track hopper located underneath. From this point, the coal fed into a crusher where it was pulverized. The crushed coal then moved by a conveyor belt fitted with a magnetic separator. This removed metal particles before it was lifted by a skip hoist into storage bins in the boiler house. These bins had a total storage capacity of 900 tons. In addition, ground storage space was available in the powerhouse yard for an additional 2000 tons from which coal could be reclaimed by mechanical means.

For steam production, coal was fed into a total of five horizontal water-tube 1521 h.p. Springfield cross-drum boilers, each working at 325 lb. per sq. in. The process began with crushed coal dropped from the bunkers by gravity through spouts into pulverizing mills that were located on the ground floor. The mill ground the coal to a fine powder that enabled it to pass through a 150-mesh screen. The coal was then delivered by air pressure through a pipe to a cyclone separator. This cyclone was located at the top of the building and any remaining larger particulate was filtered out there. From the cyclone, the coal powder dropped into a screw-operated horizontal conveyor. This conveyor moved coal to a 130-ton storage bunker located above the firing floor. One 130-ton capacity bunker was provided for each of the five boilers and was capable of holding one day's supply of fuel. Coal powder was moved from the bunkers by screw feeders to ten jet burners in each boiler. These jets were mounted to

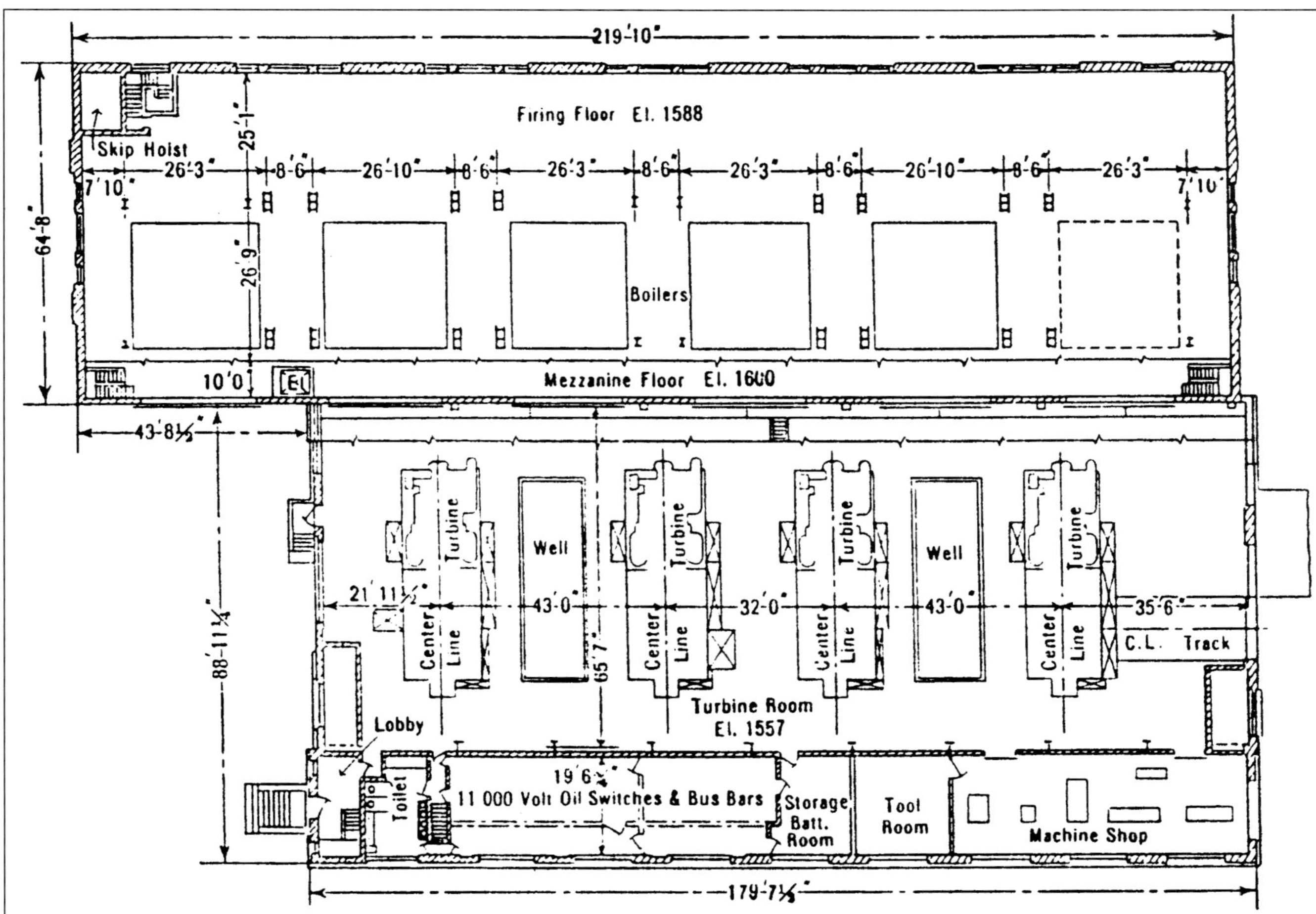

The floor plan of the turbine and firing room shows the arrangement and layout of equipment. *N&WHS Archives*

inject the fuel into the furnace in divided and nearly vertical streams.

Pressurized air was used to inject the coal into the boiler, acting to disperse the fuel at the proper velocity. A certain amount of air was drawn in around the coal nozzles, but the main amount was shot through inlets in the firebox wall. Because of the excessive temperatures that built up in the brickwork, the furnace side walls were hollow. Air for combustion passed through these spaces and in the process was pre-heated.

The water-tube type boiler was arranged so that a certain amount of tubular heating surface was provided in the water screens across the bottom and rear wall. This screening served to protect the lining from excessive temperatures by keeping combustible material away from where the jets impinged the walls. The tubes at the bottom were also protected against buildup of slag, causing the ash to form small particles, which were quickly removed by water flowing in a sluiceway. Controlling the amount of fuel and air to the burner regulated combustion control. Individual burners could be cut in or out by a control box at the boiler and also by switches from a central control panel. Automatic pressure sensors controlled the burners by cutting in jets as needed.

The steam produced was supplied to four 25 cycle, 3 phase, 11,000 volt turbogenerators. These had cast steel bed plates mounted on box girders that were filled with concrete and supported on cross-braced concrete piers. This construction eliminated vibration. A "surface-type" condenser was mounted underneath each turbine. These had a cooling surface of 20,000 sq. ft. A cool water supply of 20,000 gallons per minute was admitted from an intake tunnel on the upstream side of the powerhouse. The cooler temperatures from the river water converted the steam back into water in the condenser. This saved the installation of condenser pumps that N&W found necessary at Bluestone Junction. The condensed water could then be re-used for steam production and the heated river water discharged into an outflow tunnel down stream. Each generator was capable of producing 15,000 k.w.of power, with a continuous rating of 10,000 k.w.at 80% power.

The power requirements for the whole railroad could be handled by three turbogenerators with the fourth serving as standby should repairs become necessary on one of the units. From the generators, the current passed through disconnecting knife switches to three single-phase 10,000 k.w. step-up transformers. These transformers raised the potential to 88,000 volts. The current then passed through oil-circuit breakers to the high-tension line and the sub-stations along the line. A fourth step-up transformer was installed in 1928 after the system was finished in order to increase capacity. A variable water rheostat arrangement, similar to N&W's Bluestone installation, absorbed excess power that was returned to the plant through regenerative train braking.

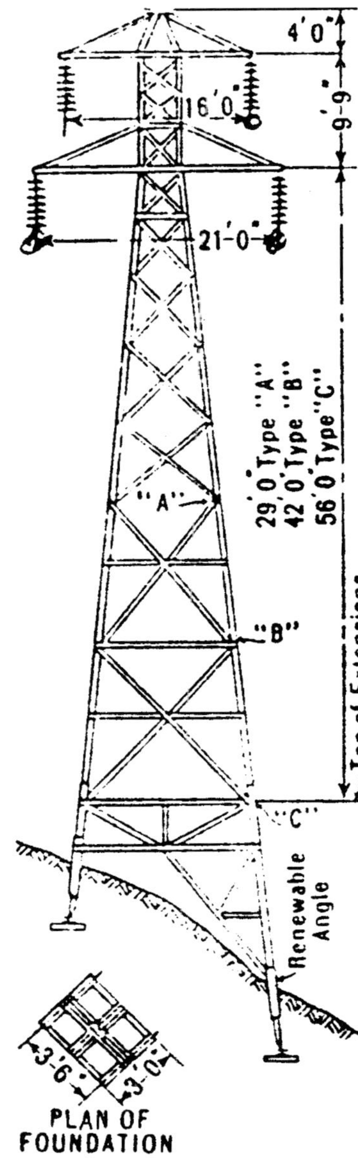

The design of the 88,000 volt suspension tower was rugged; it transversed difficult terrain away from the railroad right-of-way. *N&WHS Archives*

Construction work proceeded on the powerhouse, high tension line and catenary simultaneously. The high tension transmission line went cross-country resulting in a 34 mile savings over the railroad right of way. In the coal region, this high-tension line was carried for an additional 22-miles on the catenary. At Matoaka, WV a 9-mile high-tension transmission line tied the system into N&W's lines and Bluestone powerhouse. The high-tension transmission line consisted of 377 towers and 332 miles of conductor cable. This cable contained a total of 530 tons of copper. The towers carried two single-phase 88,000-volt circuits, with the wires arranged vertically, two on each side. On the upper cross-arms, the wires were spaced 16 ft. apart horizontally while those on the lower arm were 21 ft. apart to prevent the possibility of wind whipping and short-circuit. East of the Narrows powerhouse, the conductor was size 3/0 hard drawn stranded copper. To the west size 4/0 hard drawn stranded copper was used. A telephone circuit was added to the existing line side wood pole telephone and telegraph lines for patrol use. For the cross-country transmission lines, laterals were run at intervals of about three miles from the right of way.

Seven step-down transformer stations were installed, with one each located at Elmore, WV, Clark's Gap, Princeton, Narrows, VA, Eggleston, Merrimac and Wabun. These functioned to step-down the 88,000 volts from the transmission line to 11,000 volts for the trolley. It also provided for switching and sectionalizing of the catenary. The sub-stations housed no equipment that required a regular attendant. The transformers were self-cooled and the other switches, lightening arrestors and apparatus needed no housing. All of these substations were situated near a telegraph office or signal tower, where an existing employee was available in case it became necessary to open or close the oil-circuit breakers controlling the substations. This was done by telephone instruction from the power director.

A unique part of the Virginian system, and a way in which it differed from the N&W electrification, was the use of a "balancing wire." The balancing wire eliminated induction interference. This system was an integrated network, including switching facilities, transformers, catenary-supported contact, balancing wires over the tracks, and track bonding and return conductors.

In this system, the balancing wire and balancing transformers were the key feature. The arrangement of the transformers allowed a maximum separation of distance between existing step-down substations and at the same time reduced induction in near-by communication circuits. There were seven balancing transformer stations located midway between the main power substations. These balancing substations were installed at Herndon, WV, Rock, Kellysville, Norcross, VA, Whitethorne, Fagg, and Roanoke. They had no connection with the 88,000-volt transmission line, and were not considered "substations" in the ordinary sense of the word. But because they contained the same autotransformers, non-automatic disconnect switches, and lightning arrestors, they came to be known as "balancing transformers" by the personnel. To provide electricity for the "balancing" system, a third coil tap was taken from the low tension side of the transformer in these substations. This furnished current of double the

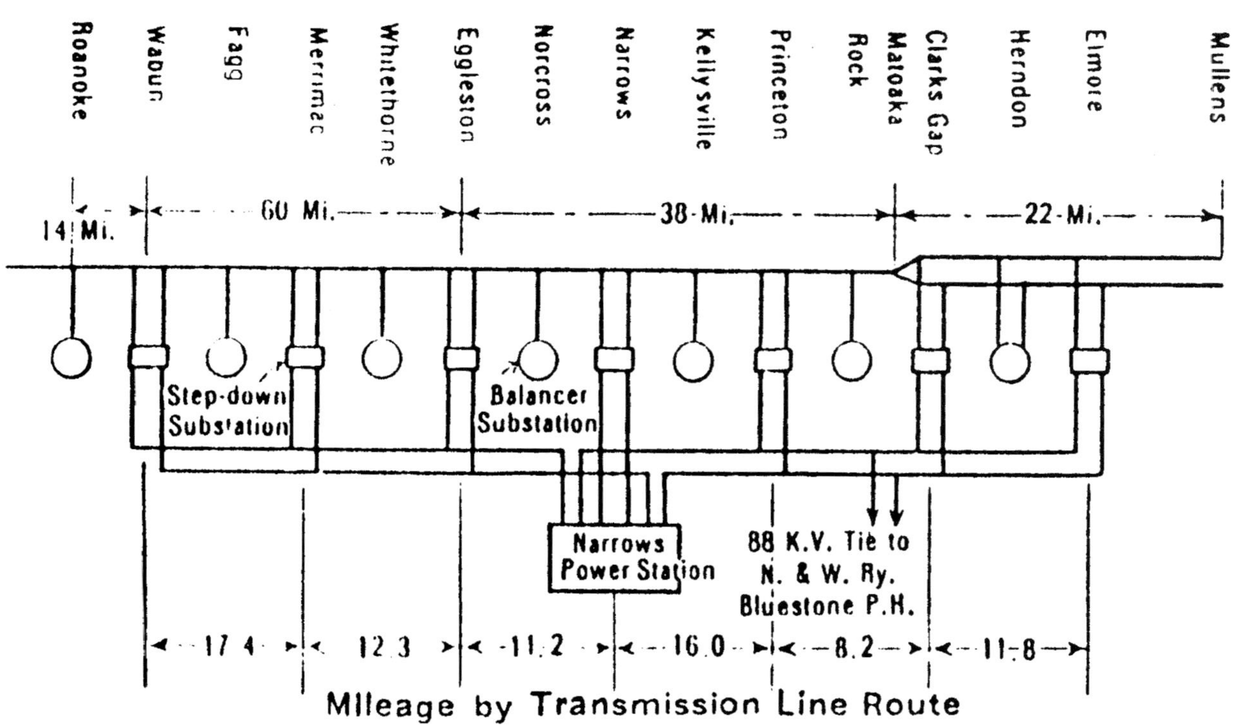

The power distribution system included both step-down transformer stations and balancing transformers.

N&WHS Archives

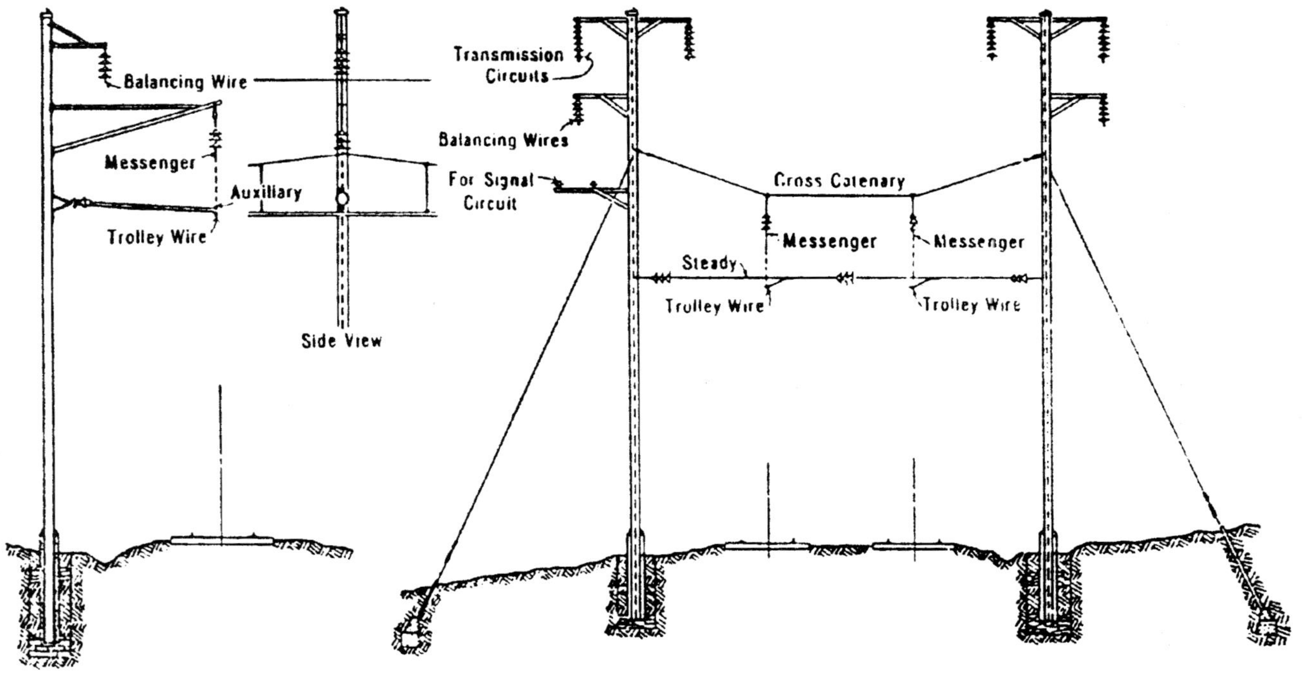

The catenary differed in the method of suspension, depending on whether it was a single or double track location.
N&WHS Archives

11,000 trolley voltage to a balancing wire at reverse polarity, which was carried on the catenary poles. It connected with all main sub-stations and the auto-transformers.

The three-coil step-down transformer at the step-down substation had one of its secondary windings connected between the trolley wire and the track. This furnished propulsion current at 11,000 volts. The other secondary winding was connected between the balancing wire and the track. This carried a balancing current of 22,000 volts. The trolley wire, track and balancing wire thus formed a three wire system having a connecting feeder of double the trolley voltage. There were three advantages gained: (1) the higher auxiliary transmission voltage of 36,000 volts resulted in an economy of copper which permitted a greater through feed between sub-stations; (2) the contemplated ultimate use of 22,000 volts at the trolley wire was already counterbalanced by the opposite polarity of the balancing circuit; and (3) better balancing of track circuits to points of load, thereby reduced the existing induction effect on adjacent communication circuits. The balancing wire in effect formed an auxiliary transmission circuit, albeit of opposite polarity through which it reduced induction interference. This permitted the installation of an auto-transformer at the far end of the railway electrification without the need of extending the high tension transmission line an additional 11 miles. It also allowed the step-down sub-stations to be placed further apart resulting in additional savings.

In the event of a derailment that caused a short-circuit in a trolley wire section supplied by two substations, the circuit breaker on the substation supplying the heavier current caused by the "short" opened its circuit breaker first. This left a stub-end short circuit flow of current from the other substation until the current was interrupted by the circuit breaker at that point. In the interim, severe induction interference would occur in a two-wire system (such as the N&W system) making line-side communication circuits unusable. In the Virginian's "balancing wire" system, the opening of one high-tension circuit breaker and the bus tie-in breaker left both secondary coils of the step-down transformer connected across the trolley and the balancing wires because the automatic circuit breaker was connected on the high tension side of the transformer. This acted as an auto-transformer, so that stub-end induction effects on communication circuits were avoided at all times.

Virginian's catenary differed from that of N&W as well, reflecting improvements. All supporting poles were single piece, brackets were cross-wired to support the catenary proper, and back-guying was used to provide anchorage against cross-stress. The trolley and supporting wire varied on different parts of the system. N&W had used the same size wire throughout, and had to deal with overheating on heavy grades. This overheating resulted in lost tension and sagging in the wire, which required the addition of a second auxiliary messenger wire to correct between Maybeury and

Elkhorn Tunnel. On portions of the Virginian right of way where heavy demand was expected (i.e. Elmore to Clark's Gap) the wire was designed to carry 1500 amperes for 10 minutes on a hot day. On the lighter grade portions, the system was designed to carry 850 amperes for 10 minutes under similar conditions.

The trolley wire was size 3/0 grooved wire of copper tin bronze with a tensile strength of 60,000 lb. per sq. in., and a conductivity of 65% of pure copper. It was designed to withstand high tensions without permanent stretch. The auxiliary messenger wire was located directly above and parallel to the trolley and was made of hard drawn copper. This acted as an electrical feeder and also supported the trolley by use of clips. Above these two wires was the messenger wire, composed of a stranded bronze core of 125,000 lbs. per sq. in. tensile strength and surrounded by a covering of hard-drawn copper. No special feeder cables were used, although the balancing wire was a feeder, in effect used for a different purpose.

The trolley, which made contact with the locomotive pantograph, was carried from the auxiliary messenger by clips, and this in turn was suspended from the messenger by wire hangers that were clipped to the auxiliary wire. The messenger wire was hung from a suspension insulator by a bracket carried on a side pole on single track applications. The messenger was suspended from a cross catenary wire that ran between two poles on multiple track settings.

Special methods were used on curves, switches and section breaks. Like the N&W, the Virginian used "chord" construction, (a series of pull-offs used to keep the trolley and messenger wires near the center line of track in curves). Virginian differed in that it used this application only in yards. On main line applications the railroad used "curve catenary" construction. This system used inclined hangers hung from the messenger wire whose positioning was determined by calculations. This resulted in a curving catenary and was useable in practically all applications. In exceptionally sharp curves, pull-offs could still be used if necessary. Virginian catenary presented a cleaner and neat appearance.

Another way by which Virginian catenary differed from N&W design was section breaks. N&W used wooden sticks, which in time were found to be less than satisfactory. They were mechanically uncertain, varied in strength according to the defect in the wood, and were so heavy as to cause hard spots in the wire, which could cause arcs as the locomotive pantograph passed. Instead of using wood, Virginian employed "air-gap" section breaks. In the air-gap design, two different sections of trolley gap wire lapped past each other, separated horizontally by about 12 inches. At the point where a locomotive pantograph transferred from one trolley wire to the other, the wires gradually rose above the level of the contact line. The lapped trolleys were suspended from the same messenger wire by hangers and were kept horizontally separated by bronze deflector rods. Insulators used to attach the trolley were three-bell suspension porcelain types, 10 in. in diameter. They were tested to withstand a dry flashover of 190,000 volts. On account of clearance in some tunnels, suspension insulators could not always be used. In these special locations, a two-piece rigid type of insulator was installed in series.

Supporting structures were steel poles, placed 300-320 ft. apart on tangents. On curves, the spans were shorter. Virginian selected the H-column structure for its catenary poles because maintenance workers using climbing skates could easily climb them. By using these skates, wire attachments and modifications were easily made. Single-piece columns were specified over less expensive lattice construction. It had been learned on the N&W installation that dirt and moisture collected at joints, leading to corrosion.

The means of returning current from the locomotive to sub-stations was through bonded track rails. Virginian used both 130 lb. and 100 lb. rail on its mainline with lighter rail in sidings and yards. All main running tracks had two bonds per joint while yards were single bonded, on one rail only. Track bonds were size No. 1/0 copper with expanded pin terminals, installed under the joint plates. Cross-bonds of one size 4/0 copper were inserted at 1/2 mile intervals. At the main sub-stations and balancing transformer sub-stations the track rails were connected to the buses by weather proof insulated copper cables laid underground. This completed the electrical circuit and allowed the system to operate.

The number of locomotives needed was determined from looking at train sheets showing projected traffic. Allowances were made for unexpected delays, flexibility was planned for meeting business variations, and extra locomotives allotted to allow for shopping purposes. For purposes of comparison, it was found that 64 Mallet locomotives would be required to handle the traffic. Construction began on the first of 36 locomotives by builder Alco/Westinghouse in April of 1925.

Mechanically and electrically, the Virginian locomotives were near duplicates of N&W's split-phase Class LC-2 engines, delivered in 1924. Thirty of the engines were set up as 3-unit locomotives with a 2-8-2+2-8-2+2-8-2 wheel arrangement. The 3-unit sets were assigned road numbers 100-109 and classified EL-3A. Six locomotives were set up to operate as single locomotives, assigned road numbers 111-115, and classified EL-1A. These single unit motors were at certain times combined to form additional 3-unit locomotives, then broken up again depending on power requirements. This 3-cab locomotive on the Virginian represented the greatest concentration of tractive effort in one singly-controlled machine found on any contemporary railroad, steam or electric.

In determining the type of locomotive much work went into examining the difficulty of starting long and heavy trains on mountain grades. A successful start could only be made gradually in order to avoid "break-in-twos" from start action. On heavy grades, a pusher was required on the rear of the train, which had to come into action at the correct time to take up its share

Due to the difficulty of the terrain, Gibbs & Hill determined the Virginian would need 3-unit locomotives, similar to models that had been recently delivered to the N&W.
N&WHS Archives

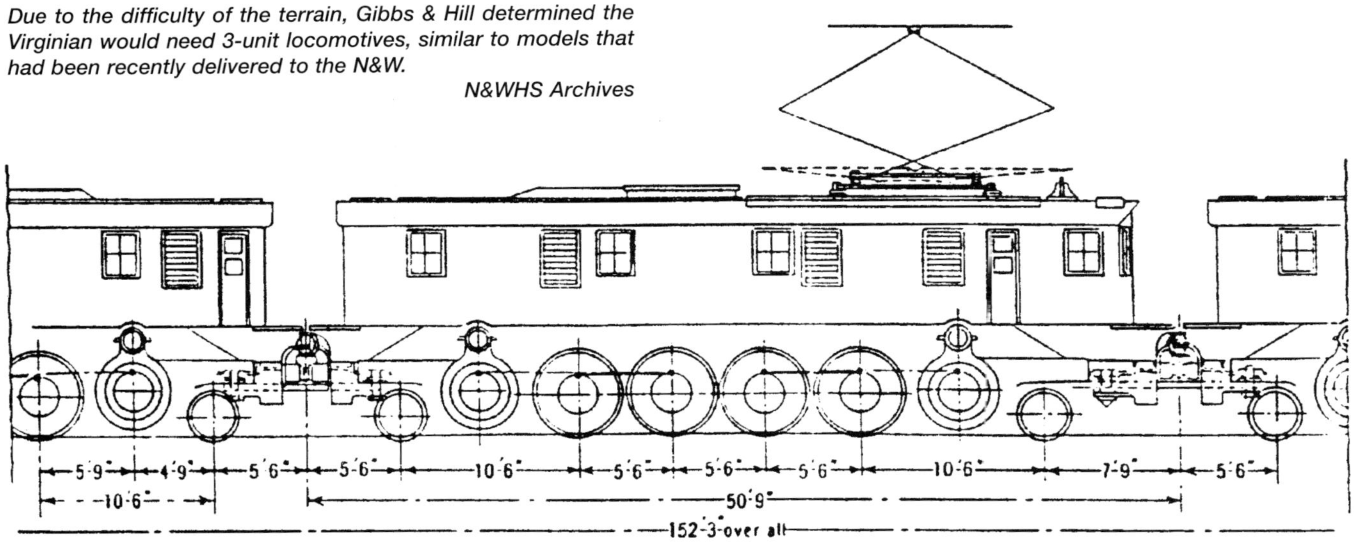

GENERAL DATA FOR THREE-UNIT ELECTRIC LOCOMOTIVE, VIRGINIAN RAILWAY.

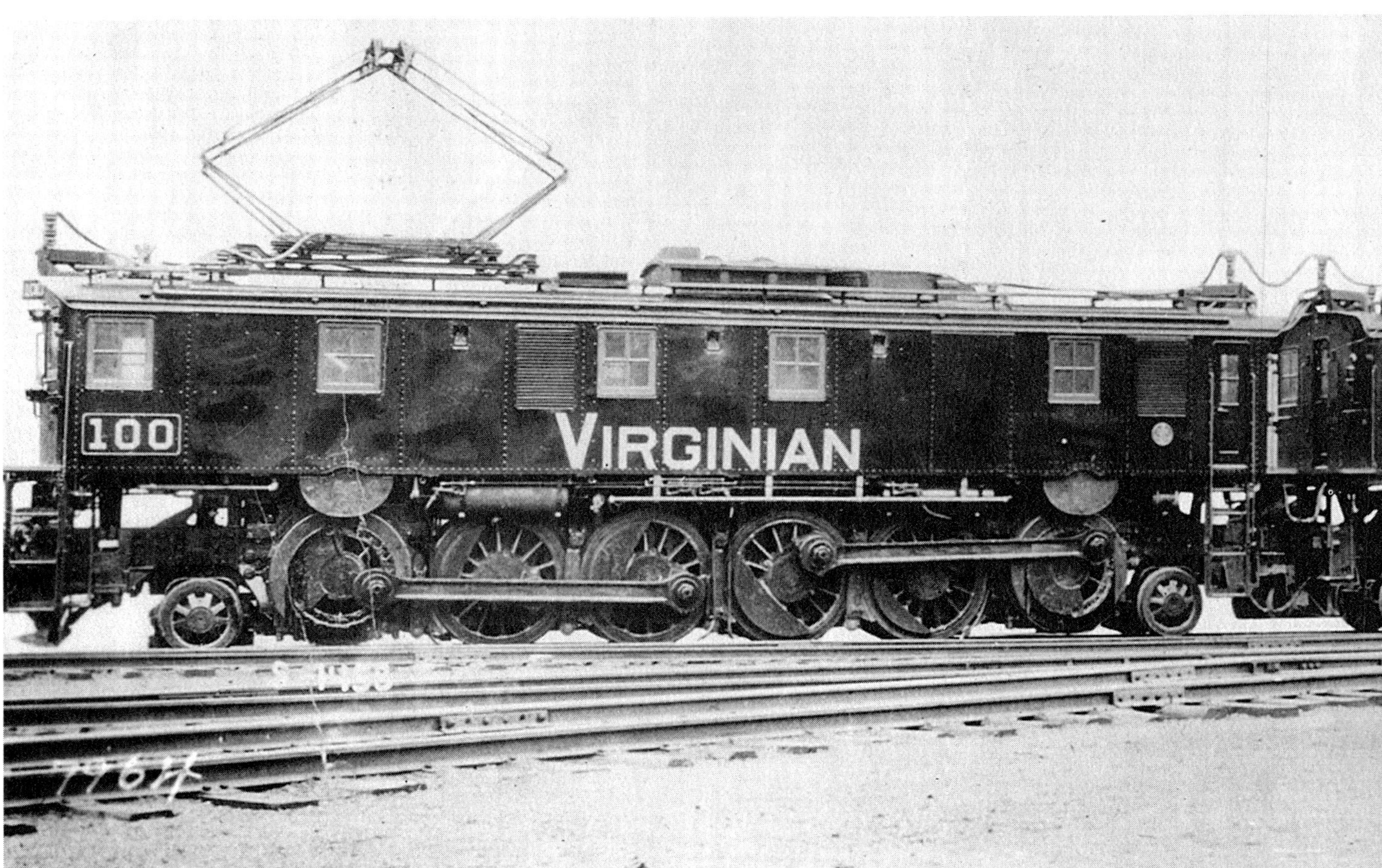

of the load. Due to the hills and curves on the western part of the Virginian, there was only one place where the rear end of the train could be seen from the front. It was difficult to start these moves even with steam, as the sound of whistles would not carry in these steep valleys. It was necessary for the head locomotive to push slack back to give a "bump" signal to the pusher to bring it into action. Often a considerable amount of time would elapse in obtaining concerted activity from both ends of the train.

Virginian planned to retain this type of operation with electric traction. Under these conditions, the motors were subjected to high current inputs and certain types of locomotives could be damaged. The split phase engine met this requirement as it was able to withstand an overload better on account of their simplicity and ruggedness. A one-third increase in tractive effort of the locomotives was achieved by adding another unit if necessary. Provisions for this contingency were made, not only with the locomotives, but throughout the entire electrical system.

The locomotives were almost identical in design to the N&W Class LC-2 engines. They received power by the locomotive pantograph, which had lateral movement to account for variable wire heights. Power was passed through an oil-type circuit breaker to a single oil cooled transformer where it was stepped down to a low voltage for the motor. The oil-cooled transformer had its windings and core immersed in a tank containing oil for cooling purposes. The oil was constantly circulated by pumps through a radiator against which air was directed by a motor-driven fan for cooling purposes. The single-phase current passed to a phase converter which converted into three-phase power for the induction motors. The conversion took place by means of a rotary "phase converter". This device consisted of a big synchronous motor with a two-phase wound stator and a wound rotor. Power was fed into and connected in the same manner as the N&W engines so that three-phase power was generated. Gradual acceleration of the locomotive was accomplished by inserting a variable resistance into the secondary circuit of the traction motor by means of a electrolytic rheostat. Power was supplied to two Westinghouse model 452-A induction-type traction motors, one set in each end of the frame and geared to a jack shaft mounted directly beneath. The motors were three-phase, having non-variable shaft speed characteristics meaning they could start from a stop and accelerate up to two certain set speeds. They were able to do this irrespective of the load. The speed was determined by the combination of motor pole connections set up. The split-phase engines

Except for minor variations, the Virginian EL-A3 locomotives were nearly identical to N&W's Class LC-2s. They were coupled in sets of three units to cope with the more severe 2.11% grade to Clark's Gap.

operated at two road speeds, 14 and 28 m.p.h. The 14 m.p.h. setting was used on the heavy grades and the higher speed was used on level and light grades. An EL-3A could exert 277,500 lbs. of tractive effort and 7,125 h.p. at 28 m.p.h.

Locomotive inspection and repair facilities were required for the electric locomotive fleet. These provisions included a track pit with housing over it, sanding facilities, and a supply of water to be used for refilling the locomotive control rheostats. These facilities were required to be located for quick access. At Roanoke, it was determined to be undesirable to hold locomotives out of service for a considerable length of time. Only routine inspection and minor work was to be done there. The main repair shops for steam locomotives and cars, as well as the Motive Department headquarters were at Princeton. It was deemed inexpedient to depend on Princeton for electric overhauling and repairs, as it was not conveniently located to the point where they were held for train service. That point was at the western end of the electric zone in the coalfield. After much consideration, they decided to use the existing Princeton Shops only for the heavy and class repairs of the internal electric equipment. For general overhauling, minor repairs and inspection, it was decided to build a new shop at Mullens WV.

The main building at Mullens Shop was a steel frame and brick structure housing a machine shop and a three-track inspection bay. The machine shop was a 316 ft. by 40 ft. facility, with the front 50 ft. area fitted for offices. A mezzanine overhead was used for light machine work. The rear 60 feet housed a blacksmith shop, foundry and a welding shop. A 15-ton traveling crane spanned the shop floor, traveling the floor length of the building. A lean-to located alongside the machine shop housed a carpenter shop, a dip-room, and two-ovens used for impregnating and baking of motor coils. These ovens were electrically heated and equipped with automatic temperature controls and recorders. The locomotive inspection portion of the structure was 206 ft. by 70 ft., covering all three tracks. An inspection pit ran the full length of each track, with a cross pit at each end for dropping wheels. These cross pits were equipped with hydraulic jacks. A 30-ton electrically driven traveling crane with a 10-ton auxiliary hoist spanned the main shed.

From the moment the decision was made to electrify in the spring of 1923, the railroad made steady progress. Construction began on the power plant at Narrows and on catenary erection along the right of way. Sufficient progress had been made by September 14, 1925 to energize the 14-mile section from Elmore to Clark's Gap. A 20.5-mile extension was opened to Princeton on October 1, 1925, and immediately placed into operation. Further work was delayed pending on going merger talks with N&W. Public

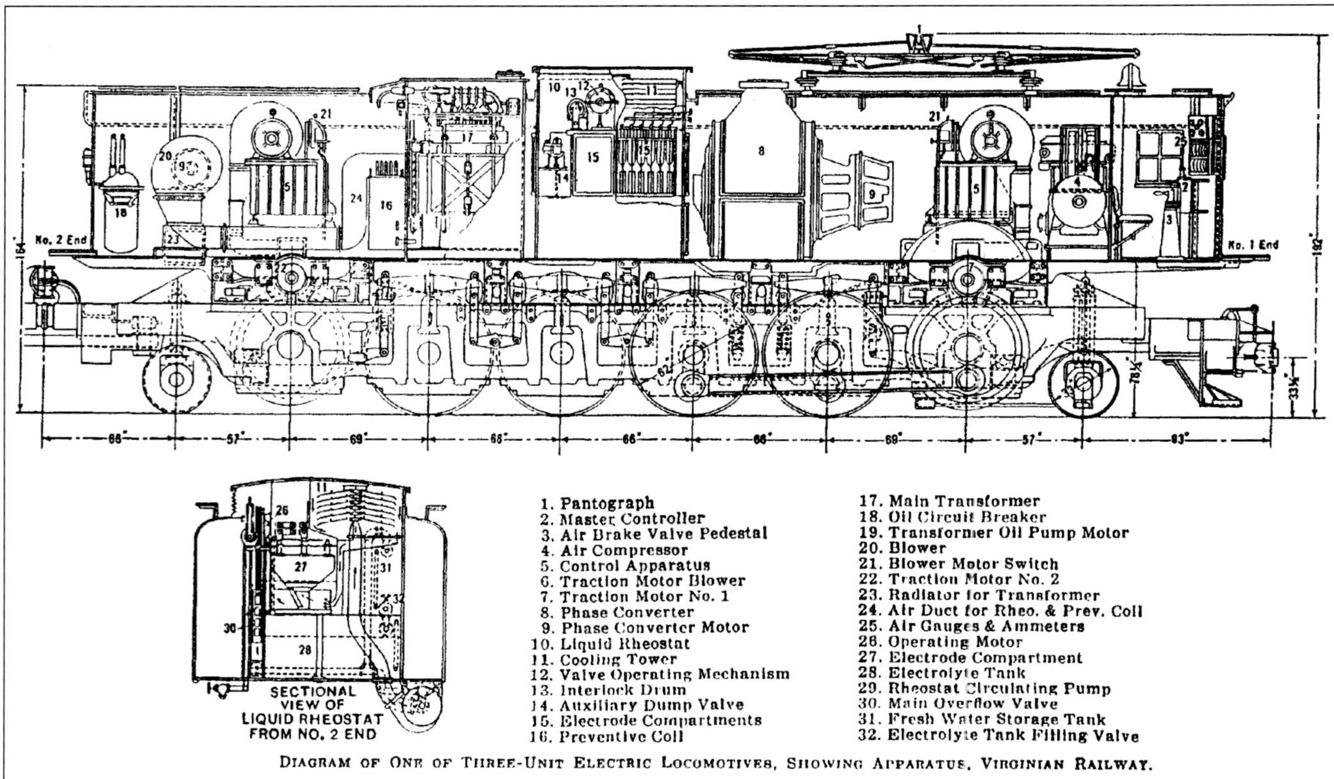

The layout of a Virginian EL-3A locomotive unit duplicated the N&W Class LC-2 engine.

N&WHS Archives

announcement of a Virginian lease by N&W was made on April 22, 1925. The following month stockholders on both roads approved the deal. Application was made to the I.C.C. on July 2nd, with C&O intervening to lead the opposition. In testimony before the I.C.C., N&W's future plans for the Virginian electrification were revealed. N&W President Jenks testified $10 million could be saved by utilizing Virginian's power station at Narrows and electric locomotive fleet rather than spending the money to electrify the N&W mainline between Iaeger and Williamson. This switch would release 36 mallet type locomotives, in service at that time on the section, for other service for an estimated saving of $2 million dollars. Additional savings were promised but a large number of persons opposed the application, mostly on grounds of lessened competition. The I.C.C. agreed and the lease was disapproved. Work on the Virginian electrification continued with the first electric train operating into Roanoke on September 18, 1926.

The Virginian, justifiably proud of its new electric operation, arranged an exhibition to celebrate its opening. Leaders of industry and the press were invited to attend. A 62 car, 6050-ton coal train made up of 50 and 116 ton cars was assembled for the celebration. A caboose and the Virginian business car were tacked on the rear, then two of the 3-unit EL-3A electrics, one pulling and one pushing, were added to the consist. This brought total train tonnage to over 8000. In order to provide a comparison for the skeptics, a similar coal train was assembled with 3 mallets providing the power. The steam-powered train was given a 15-minute head start up the double tracked grade to Clark's Gap. The electrically-powered train overtook the steam-powered run before they reached the mid-point of the hill. Here, batteries of newsreel and movie cameramen were on hand to capture the event. Time up the 14-mile grade to Clark's Gap was made in 1 hour with the electric run. The trailing steam train although slightly lighter took 2-1/2 hours to cover the same distance. Finishing touches to the project were added over the next few months and electric train serv-

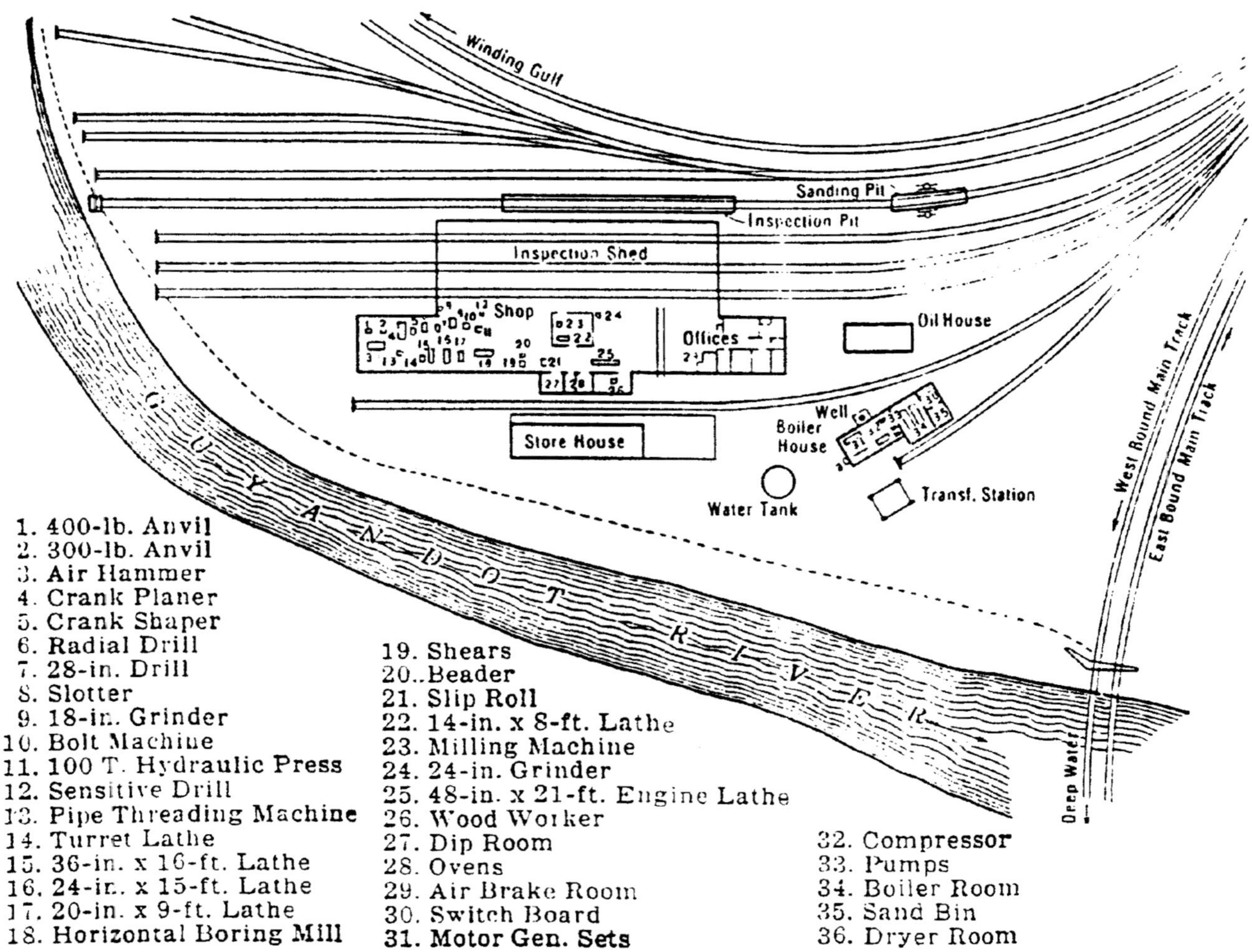

LAYOUT OF MULLENS SHOP, VIRGINIAN RAILWAY.

Virginian's maintenance facility was located close to the source of traffic so that units could be more readily called to duty.
N&WHS Archives

ice settled into a regular pattern.

True to expectations, tonnage increased from 7,440,832 tons in 1924, the last year for exclusive use of steam motive power, to 11,825,101 tons by 1927. This was due for the most part to an increase in capacity, which resulted from a general "speeding up" of train operations by the introduction of electric power. The electric engines could move a train in less than half the time of steam. The Virginian settled into a routine electric operation in which a 6,000-ton coal train was assembled in the yard at Elmore WV; it was given 2 of the 3-unit EL-3A electric locomotives, with one set pulling and the second set pushing. This train, known as a "Hill Job," did not operate all the way through to the next terminal. At Algonquin, on the east side of the Clark's Gap tunnel, the coal loads from the "Hill Job" were set off on a storage siding and the motive power returned to Elmore. A following 6,000-ton train was assembled at Elmore, again with 2 of the 3-unit EL-3A electric engines. Again, one set of engines pulled while the second set of locomotives pushed. The road train climbed the grade to Clark's Gap, then filled out from the siding at Algonquin with approximately one-half (3,000 tons) from the preceding "Hill Job". Now with 9,000 total tons, the train moved off for Princeton with one 3-unit EL-3A locomotive. The pusher cut off at Clark's Gap and returned to the yard at Elmore.

Traffic dropped significantly during the depression. The economies of electric service kept the fleet in operation, while the use of older steam power was sharply curtailed. Coal traffic rebounded to over 18 million tons during WWII. At this time, expenses began to rise on the electric operation. By the mid-1940s it had been in continuous use for twenty years. The system was nearing the end of its useful economic life. Several improvements were made to upgrade the Narrows powerhouse beginning in 1946. A new Foster-Wheeler boiler was installed along with a 10,000 k.w. General Electric turbogenerator. Additional equipment was added again in the years 1949 and 1950, resulting in an overall 25% increase in capacity.

To augment the aging Class EL-3A locomotives, four new locomotives were purchased from General Electric in 1948. Classified EL-2B, and assigned road numbers 125-128 by Virginian, the massive engines featured a diesel-styled carbody and operated in pairs. They were quickly nicknamed "streamliners" by Virginian personnel. Weighing nearly 517 tons, press releases repeatedly called attention to the engine weight of over 1 million pounds. Stretching 150 ft. 8 in. between couplers, the pair developed 6800 h.p. with a short term tractive effort rating of 260,000 lbs. More significantly, the EL-2B could produce a continuous output of 162,000 lbs. tractive effort, making them again the most powerful continuously-rated locomotives in the world. Unlike the EL-3A engines, which featured a split-phase converter, the EL-2B locomotives operated on a motor-generator system similar to units General Electric had built for Great Northern. Each engine had a transformer that reduced incoming 11,000 volt A.C. to operate a synchronous motor. This motor drove a D.C. generator and a D.C. exciter. The locomotives rode on a B-B+B-B wheel arrangement and were equipped with the proven General Electric model 746 traction motors in a nose-suspended diesel design. Each traction motor was equipped with a pair of the 42 in. diameter driving wheels.

The powerful locomotives were an immediate help to operations. The builder had promoted the new engines for heavy mountain work for other railroads. In one particular test by General Electric, one set of EL-2B engines took a 3,000-ton train up a 1.3% grade at 35 M.P.H., and rolled a 10,000-ton train on level track at even higher speeds. In regenerative braking, the EL-2B could generate nearly 7900 braking horsepower, which was more than sufficient to control heavy trains on a grade without car brakes. The EL-2Bs had a maximum speed of 50 m.p.h., significantly faster than the 28 m.p.h top speed of the EL-3A engines.

A hard look was taken at electric operations in 1950, at which time the railroad was dieselizing other portions of its rail operations. The EL-3A Class engines were worn, needing an inordinate amount of attention to remain roadworthy. Locomotives from this group were being retired on an independent basis depending on condition and cost of repairs. Virginian decided to retain its electric operations based in large part on its recent investments, which had lowered operating costs. Powerhouse improvements and the EL-2B engines, for which the railroad had calculated maintenance charges of only 5 cents per thousand gross ton-miles, played a decisive part in the choice not to dieselize.

The Virginian was interested in new advances in electric traction design that appeared in contemporary journals, particularly the ignitron-rectifier concept. Both the Pennsylvania and the New Haven Railroads were experimenting with this system on several existing passenger locomotives. After further testing and improvements, General Electric began delivering the first of 12 new units based on this design to the Virginian in 1956. These locomotives were equipped with a Pyranol-cooled main transformer and 12 liquid cooled mercury arc ignitron-rectifier tubes. These tubes converted 11,000 volts A. C. received from the catenary to D. C. for the traction motors. The ignitron-rectifier was a simple design and easy to repair. The locomotives had many of the features of diesel-electrics including the ability to M.U. and two 6-wheel trucks.

Virginian specified that the new electric locomotives accept FM trucks so it could use its spare diesel set during electric motor truck overhauls. The 69 ft., 6 in. engines weighed 174 tons. Classified EL-C, each locomotive developed 3300 h.p. and 98,500 lbs. tractive effort. By early 1957, a total of twelve EL-Cs were delivered and assigned road numbers 130-141. During break-in tests that spring, a pair of EL-Cs took a 3140 ton coal train up the 2.11% grade to Clark's Gap, where it filled out to a total of 10,000 tons before continuing to Roanoke unassisted. This performance allowed the high maintenance 1925-era EL-3A engines to be pulled from road service, although many remained stored on the property. The only mechanical problems that occurred with the EL-C Class resulted from buffeting from slack action, which caused the ignitron-rectifier tubes to leak. By the end of the decade, Virginian had rehabilitated its electric operation by increasing both the capacity of its power plant and modernizing the motive power. Virginian now had the most modern electric operation in heavy-duty railroading.

N&W had attempted merger with Virginian in 1925, only to be rebuffed before the I.C.C. In a second attempt in 1959, the railroad was successful, benefitting from a changed attitude. Soon the new Norfolk & Western was busy implementing changes to bring about operating economies. The merger changed the flow of rail traffic in several important ways. On the eastern end of the Virginian, through coal trains to Norfolk were routed over N&W tracks at Abilene VA. Virginian track to Jarrett was retained for local service only, along with some lines in the Norfolk area. Heavy coal trains were diverted off N&W's Blue Ridge crossing to take advantage of the nearly gradeless Virginian line along the Roanoke River. Between Kellysville, WV and Roanoke, VA, the eastbound routes of both roads were combined with coal traffic using the easier Virginian for the Alleghany crossing in the Christiansburg/Merrimac area. On the western end of the former Virginian the bulk of coal traffic was redirected to follow the Guyandot River west from Elmore to a N&W connection at Gilbert. Here it turned eastward, thereby avoiding the more severely graded Virginian line over Clark's Gap. The Virginian electric operation was now, for the most part, out of work.

Virginian modernized its electric motive power fleet after WWII with the addition of the million pound EL-2B locomotives from General Electric.

The last of the Class EL-3A boxcab locomotives remained on the property at the time of the merger, stored at Roanoke and Princeton. By 1960, the last of these 1925 era

The Virginian EL-C engines were capable of sustaining electric operation well into the future. In this photo, taken at Princeton, WV after the N&W merger, an east-bound train shows one immediate change under N&W: the use of three locomotives per train instead of two.

EL-C No. 235 was the only member of the Class to receive N&W paint. This was the result of repairs to wreck damage that were not completed until after the merger.

motors went to the scrap lines. The balance of the Virginian electric locomotives were then re-numbered into the N&W motive power roster. The four 1948 EL-2B sets, (Virginian road numbers 125-128), were split up, with each of the pair given its own specific N&W road number, 221-228. The EL-C class was re-numbered to N&W road numbers 230-241. Of this group, only the 235 was repainted into N&W colors. EL-C 235 featured new black body paint with dulux gold lettering. The repainting resulted from a sideswipe accident at Kumis, VA. The "then" Virginian EL-C 135 ended up in the Roanoke River and was still under repair at the time of the merger. Electric operations remained in place on the former Virginian route from Mullens to Roanoke. N&W operated the EL-C engines in groups of three instead of two, as per former Virginian practice. This was in part due to increased train lengths with Norfolk & Western now running the road.

After the merger, the EL-2B electrics continued to work out of the former Virginian coalfields. Passing the station platform at Mullens, WV, we see one of the changes that have occurred: the influx of N&W hoppers in the yard.

ELC-235, in N&W paint, lays over at Mullens, WV. The photographer has cleaned the accumulated road grime off the nose herald for his shot.

Another change post-merger: the breaking of the EL-2B pairs of electrics, with each assigned its own road number. This seldom-seen view of the rear end of No. 221 was taken at Mullens, WV.

The electric locomotives soon clashed with the new operating plan between Kellysville, WV and Roanoke. The routing of eastbound traffic was combined between these points with loaded coal taking advantage of the better grades on the former Virginian lines and empties returning via an N&W route. The electric engines worked eastward with loads, but since there was no catenary on the N&W side they had to return via the Virginian against the flow of traffic. Norfolk & Western was now faced with a difficult choice. The company had the alternative of expanding electrification to include both routes between Roanoke and Kellysville. This action would have allowed free movement of both loads and empties. The other option was to convert entirely to diesel and shut down the electric operation. After considering costs, the decision was made to terminate all electric operations.

An N&W coal train climbs toward Clark's Gap in former Virginian territory on June 20, 1962. Longer trains led the railroad to assign 3 units to the head end for power.

A pair of EL-2Bs shove an N&W coal train up the hill near Clark's Gap on June 20, 1962. In a few days, the electric operation would be history.

On June 30, 1962 the final electric run tied up in the yard at Roanoke. The boiler fires were killed in the power plant at Narrows. N&W electric operations came to an end. The remaining locomotives were gathered at Roanoke and advertised for sale. The unexpired equipment trust plates from the EL-Cs were transferred to newly arrived N&W GP30 diesels, technically making them the last "Virginian" locomotives. The older EL-2Bs lasted into 1963. When no offers for their services arrived, they were cut up for scrap. The almost new EL-C rectifiers sat idle for more than a year. New Haven Railroad acquired the rectifiers for the bargain price of $300,000 for all twelve plus spare parts. Soon, all except EL-C 230, which was rebuilt into a diesel hump trailer, were in regular service operating far from the mountains and hauling freight between New York and Cedar Hill yard in New Haven, Connecticut.

The rich history of electric operations on the N&W had, for the second and final time come to an end. From a beginning shortly before World War I, electrification on Norfolk & Western helped to unclog the road's rugged Elkhorn crossing and allowed the railroad to prosper. Later, a newly merged and different Norfolk & Western inherited one of the best contemporary electric operations in the land. It was cut short only by a change in operating routines. Still, Norfolk & Western had met one of the toughest operating challenges faced by electric railroading. ■

Epilogue

Riding the Electrics Through the Coal Fields

One set of LC-2 or LC-1 electric locomotives would normally be available for the westward trip into the coal fields with 80 to 100 hoppers, representing about 2500 tons. On a typical run, the crew would begin its day by registering at the call office at Bluefield Yard. The engineer and helper crew board the engines and check all gauges, handbrakes and supplies. The engineer inspects the running gear and perhaps oils around a bit. When ready, the crew eases out and backs down to the designated track according to signal to pick up its train. A car inspector couples the locomotives to the train and connects the air hose. The engineer would then pump air to charge the train line and the auxiliary reservoirs under each hopper. When fully charged, the car inspector orders an air test, and the engineer makes a service application of the air brakes to assure proper working order. The car inspector and crew walk the train inspecting each car looking for defects and proper brake application.

The conductor comes to the engine and informs the crew this train of hoppers is designated for loading with coal on the Dry Fork line. The run would work westward to Auville yard, as far as the electrification extended. The engineer will run at a speed of 28 miles per hour from Bluefield to Bluestone; 14 miles per hour up the 1% grade from Bluestone to Coaldale; approximately 20 to 25 miles per hour regenerating from Coaldale to Eckman, then 28 miles per hour to Auville yard. When receiving a signal to proceed, the engineer releases the brakes, the slack rolls out and the train begins the journey westward.

At the crest of the hill at Bluefield, WV, at MP N-363, the elevation is 2566 ft. above sea level. The train will drop as it continues westward for the next ten miles to Bluestone Junction. Standard height for the catenary trolley wire is 23 feet, 9 inches above the railhead. At certain locations, the wire is lower because of a local physical obstruction. Approaching each side of these points, where the wire was lower than standard, is a low wire warning signal. This consists of a black circle on a yellow background, illuminated by a circle of lights. Our first low wire warning is at the Mercer Street and Allen Street bridges in Bluefield Yard. We work our way out of the yard, passing the freight house on a 1.21% descending grade, passing the WV/VA state line marker near MP N-365. The train slows as it approaches the passenger station at Bluefield, VA. Straight ahead, a signal at HQ tower indicates our route is clear. The outbound hopper train works through the crossovers and through the 7 degree curve and begins the trip west.

The engineer accelerates first to 14 m.p.h., then changes to 28 m.p.h. by changing two trucks at a time. This keeps the slack in our train stretched. The engineer watches the signals as the train worms through the 6 and 7 degree curves along the Bluestone River approaching Hale. Eastbound trains are superior by timetable; this causes westbound traffic to sometimes back up on the mainline to accommodate opposing movements. We approach the crossovers at MP N-369 and pass the station at Falls Mills.

The hoppers roll west passing the Yard Limits for Flat Top Yard. On the opposite track another LC2 is filling out its coal train with additional tonnage. The train rolls over switch work and through the compound curvature at the east end of Flat Top. Some trains stop to set off cars and pick up loads from the Bluestone and Pocahontas branches. About 2500 tons of westbound loads are assembled daily at Flat Top Yard from nearby operations. Our train weaves through cuts of coal and rounds another 9 degree compound curve on mixed 130-132 lb. rail, past the Yard limit board and the Dupont Powder Company. Near MP N-372, the train passes the yard limits for Bluestone Junction.

For westbound traffic, a 3-arm interlocking signal located on the east side of the Bluestone Powerhouse controls movement through the plant. The tracks to our right hold inbound loads of coal for the power house. Electric locomotives undergoing inspection can be seen at the shop. The train rolls over a crossover, passing SU tower at the junction of the Pocahontas branch.

At MP N-374, the elevation has fallen to 2287 ft.

An engineer rolls an electric locomotive westward over the Bluestone Branch with a train of empties returning to the coal fields.

The engineer switches over to 14 mile per hour operation as the train begins it's ascent to Coaldale. The Bluestone branch diverts from a connection on the eastbound track. The train begins to climb as it works through a 7 degree curve and into the 6975 ft. Cooper Tunnel. Working out the other side and sailing over the Bluestone River, the engineer keeps a sharp eye for the signal ahead at Ruth, the end of the double track. The train slows at the approach to Elkhorn Tunnel. If congestion makes it necessary, westward trains sometimes receive a signal to cross over from the westbound to the eastbound track at the crossover located west of the tunnel at Coaldale. As a safety measure, the train line is checked by gauge in the caboose, and recorded by the brakeman before descending through the tunnel on the Elkhorn grade.

Selecting a time when the train is balanced over the summit, the engineer shuts off the throttle and switches the balancing lever from motoring to regenerating. Regenerating acts to retard movement downhill and saves wear on the brakes. Excess electricity generated by our train is returned through the catenary for other trains on the railroad. Past the congestion at the tunnel, the train drops down into the Elkhorn valley. Mining activity is everywhere with every piece of land used for some purpose. A 14 m.p.h. restriction down to and over Bridge 681 at Maybeury is strictly observed on downhill runs. The train rolls over the crossovers west of MP N-387 and increases its speed to 20-25 miles per hour. Under certain conditions, when rails are wet and slippery, it was necessary for the engineer to use the air brakes while regenerating in order to maintain control. The engineer makes a reduction of approximately 5 pounds in brake pipe pressure, which applies the brakes on the train, while at the same time keeping them released on the locomotive. Passing Lick Branch, coal mines close in on the tracks. Elkhorn Pocahontas Company is located off the eastbound main. Wyborn Coal has a loader on the westbound track.

The grade slackens to 1.4% as the hopper train winds along the valley floor. The track is rarely straight and compound curves are commonplace. PS-type (Pennsylvania Standard) rail mixes with newer types at MP N-383. The train slips past Byrd Yard, which handles loads off the Northfork branch. Towns close in on the tracks at Northfork and Keystone, along with

This is the view looking west near the set-out track passing Maybeury station.

Eastern Gas & Fuel's sprawling Keystone Operation. At MP N-388, the Yard Limit board to Eckman Yard rolls by. Westbound moves run under special timetable restriction when running the wrong main. A spring switch located at the Eckman Yard pullout has a speed restriction, which is no concern for eastbound trains leaving the facility. These movements would be below the speed applied to this type of turnout. The problem exists for a westbound train if running the wrong main against the flow of traffic. This was usually avoided unless absolutely necessary.

The train works west on the mainline, over crossovers and the spur leading southward up Dan's Branch. A tipple stands hard against the main at Landgraff. Curvature stiffens to 12 degrees as the train passes the Yard limits board for Kimball yard at Vivian. The yard used for hopper storage at nearby mines lays to the right. The engineer allows train speed to increase to nearly 28 miles per hour. A Low Wire Warning appears just ahead of the 680 ft. Kimball Tunnel. Out again into daylight, the hoppers roll west on signal indication. Right of way corrections over the years have shortened the distances between MP N-392 and N-393 to 2100 ft. The engineer keeps a lookout for traffic ahead and shifters working on sidetracks. Another Low Wire Warning appears before the train rolls through 645-ft. Big Four Tunnel on more of that P. S. rail. At MP N-395, the train blows through the short 174 ft. Big Four No. 2 Tunnel, the 362 ft. Twin Huger Tunnel and past the Lake Superior Coal tipple.

We slow as we work through a series of six-degree curves and over the crossover at MP N-398. The passenger line to Welch departs on the left. A Low Wire Warning announces the 1234-ft. Welch Tunnel. The train exits on bridge 893A, gets a roll-by inspection at Tug Tower, over another crossover, and into Hemphill Tunnels No. 1 and 2. Our hopper train gets a signal and slows approaching the five track yard and passing siding at Farm. Eastbound steam-powered trains are taking on coal and water. Coal loads off the Tug Fork branch are being prepared for the trip up the Elkhorn. The hoppers ease through the congested tracks and over a set of crossovers, past a Low Wire Warning ahead of the 613 ft. Antler Tunnel No. 1 and 2.

The Superior branch departs at MP N-406. The train passes Davy, WV then Allied Chemical's Semet Solvay plant before entering the 760 ft. Twin Branch Tunnel No. 1 on a 5-degree curve, which extends all the way through the bore. Amigo Smokeless Coal's Twin Branch tipple occupies the space between the bores before the hopper train slides into Twin Branch Tunnel No. 2 on a 2-degree curve. Marytown, WV spreads out around the right of way as we exit the tunnel. Most of the residents work at the nearby Douglas Pocahontas Coal's Marytown operation, whose tipple and load yard lay alongside the mainline. The train winds around 4 and 5-degree compound curvature past Claren, and the Beamer Pocahontas Company's tipple on the westbound main.

The Upper Hensley branch departs near MP N-411, where a set-out spur exists for the occasional hot journal box. The train crosses bridge 805 and enters the 1115 ft. Vaughn Tunnel. A crossover provides access from both tracks to the Spice Creek Branch. The train slides under the Low Wire warning and into the 924 ft. Roderfield Tunnel. Some realignment of the right-of-way has taken place over the years to ease curvature. A 6 degree curve though MP N-414 leads into Laurel Tunnel, formerly known as Boxley Tunnel No.2 on the old line. Compound 2 and 6 degree curves brings the train into Gordon Tunnel on P. S. rail. This bore, once called Serpell Tunnel, is a 1271-ft. bore and is located between MP N-415 and 416. Re-alignments have shortened the distance between these two mileposts to 1271 ft. A center track passing siding begins at MP N-417.

Wilmore, WV is a mining town housing workers at the Pocahontas Red Bird whose tipple is passed as we roll by on the westbound main. The train passes switch work for the west end of the center passing track and through the curves at Sandy Huff. Pocahontas Red Ash Mining's Pauley Operation has a large three-track tipple off the westbound track. The train rounds a 5 degree curve and passes Iaeger Pocahontas Coal's tipple on the eastbound track.

Our hopper train slows approaching the passenger

Capels, WV was typical of many of the mining towns which developed between Welch and Iaeger.

station at Iaeger. The engineer guides the train through the crossover, passes the yard board, and slows to yard speed around the 1 degree curve approaching Auville yard. The train is routed into one of the five through tracks that make up the yard. The inbound hoppers from Bluefield will be taken to mines further up on the Dry Fork line. Located at the furthermost westward assembly point of the electric operation, Auville Yard is a busy place. The yard provides service for steam-powered mine shifters working on the Dry Fork line. Besides the coal business, the line serves as a through route and is used as a shortcut for empties returning to the Clinchfield Railroad.

Here, a short eastbound coal train has been built on a track and inspected by car men prior to departure. Our run will return eastward to Eckman Yard where the coal train will fill out to 4000 tons, or about 50 cars, and pick up a helper. The set of LC-2 electric locomotives that had brought hoppers into Auville will now turn around for the 54 mile trip to Bluefield with these cars. The engineer changes operating ends on the power and opens a few windows to help dissipate heat generated by the motors. The conductor drops off and heads to the yard office to exchange paperwork while the crew eases out and backs down to the designated track to pick up its train. A car inspector couples the locomotives to the train and connects the air hose. The engineer begins to charge the train line and the auxiliary reservoirs under each hopper. When fully charged, the car inspector orders an air test, and the engineer makes a service application of the air brakes to assure proper working order. The Auville Yard car inspector and train crew walk the train inspecting each car looking for defects and proper brake application.

When finished, the conductor contacts the Power Director by company telephone before proceeding back to the train. The Power Director has the responsibility to see that enough electrical power is available for the train to proceed. A signal to proceed is received from the operator in DY tower and the engineer opens the throttle gently. The jackshaft begins to rotate, the side rods go into motion and slack is pulled out one car at a time. Our train moves through the switch work around the east leg on the wye passing the station at Iaeger.

Eastward trains were superior by timetable rule, but coal movements mixed with a variety of higher priority passenger and freight business. Delays getting started sometimes occur before a sufficient window of time exists to permit a train to depart Auville Yard. Crews usually receive a number of slow orders for track work and other cautions. Prior to departure, the engine crew reviews the train orders, but is again out on the road in the cab before a response to the train order would be necessary. Maximum permitted speed for steam-powered freight trains was 40 m.p.h. on this part of the road, while electric coal service maintained a steady 14 m.p.h. This was more than adequate to maintain sufficient separation of traffic, taking into account the need of steam locomotives to stop for water.

The train is out on the mainline and headed east. At Wilmore, a 5925 foot long eastbound siding gives the dispatcher an option for getting a passenger train around slower traffic. Should the coal train be put into the siding, speed was limited pulling through the spring switch located at the east end in the move from the siding back to the mainline. Moving at a steady 14 mile per hour pace, electric powered coal traffic maintained adequate spacing with following passenger traffic. Slower moving trains ahead were more frequently a source of delay, resulting in the occasional coal move through the siding. The large amount of opposing traffic that resulted from the war denied use of crossovers and wrong main running, which had been an option in pre-war conditions.

With the swollen number of wartime coal and freights, six normally scheduled passenger trains running each way shared the right of way through the N&W coalfields. Delays were encountered as these trains made routine station stops, especially with the off-loading of express business. In addition, there were two time freights running each way, which were in effect priority trains operating on fixed timetables. These took precedent over coal traffic. Water and coal stops made by the steam-powered runs posed additional restrictions on the steady electric train running times. Coal trains "rode the yellow"... a practice of running at reduced speed in the block behind slower movements ahead. The use of passing sidings in daily operation was avoided due to delays of getting back out into the flow of passing traffic.

The first Low Wire Warning for eastbound trains is encountered approaching Gordon and again at Laurel Tunnel. Approaching Roderfield, the coal train is now 45 miles from Bluefield. Low Wire Warnings are again encountered at Roderfield Tunnels No. 1, 2 and Vaughn. At Claren, WV, our train passes a 4430 foot long siding used by both east and westbound trains before entering Twin Branch Tunnels No. 1 and 2. These bores also display the Low Wire Warnings on account of clearance, as do Antler Tunnel No. 1 and No. 2 located further east of Davy, WV.

Near Farm, WV, a Low Wire Warning is encountered at Bridge No. 898A. Here the coal train slows. To the east, the right of way is crammed into the valley floor. Mining, housing or the railroad uses every bit of space. Located at the bottom of the steepest part of the hill, the coal trains join the company of other steam-

The Low Wire Warning signs were illuminated so that they could easily be spotted in all weather conditions. The crew could then take measures to avoid damage to the pantograph.

powered trains that are stopped to either pick up loads, or take on coal and water. A 7000 ft. passing siding eases track congestion, this being located between the east and westbound mainlines. Access is gained from an inside switch to the west end of this siding. The catenary dips again passing through Hemphill Tunnels No. 1 and 2, with the accompanying Low Wire Warnings.

The train passes Tug Tower, and gets a roll-by inspection from the operator. The passenger route via downtown Welch and the Tug Fork line up to Wilcoe Yard depart from the freight line on the right side of the train. We pass through Welch Tunnel, with it's Low Wire Warning, and exit on the east end, bypassing the downtown business district. At Huger, a 3000 ft. siding allows the dispatcher the option of getting a passenger train around slower moving traffic.

The coal train works eastward along Elkhorn Creek, passing through a series of tunnels, Twin, Big Four and Kimball, each necessary to ease curvature. A Low Wire Warning is present at each bore. A 2935 ft. siding for the eastbound mainline is passed at Kimball and the train rolls on eastward, by a coal yard at Vivian. Mining activity is more intense and present all along the line. As the train nears Eckman, the distance has closed to within 25 miles of Bluefield. Eckman is the site of a coal assembly yard, located on the right side of the passing train. The train enters the yard to pick up additional loads and a helper.

While additional loads are added to the train, the dispatcher uses the extra mainline capacity that our coal run has just released. The crew walks the train in Eckman Yard inspecting each car as a helper locomotive is attached. Freight and passenger trains roll by on the mainline. The train is filled out at both the head as well as the rear end, paying special attention to keep either one or the other locomotive coupled to the train at all times. Movement within Eckman yard is controlled by the Yardmaster. When the train is fully assembled and ready to proceed, the Yardmaster contacts the Dispatcher and obtains permission to proceed.

In starting, the lead locomotive reverses to move backward against the train. This bunches the slack against the pusher locomotive. As soon as the pusher locomotive feels the slack, the engineer on the pusher locomotive applies steadily increasing power until he has maximum accelerating current on the motors. The engineer on the head end, who feels this starting impulse, reverses his locomotives and begins to pull. The lead locomotive begins to take up the slack, car by car, until the load has been sufficiently balanced between the lead and pusher engines to easily move the train.

The locomotives are equipped with relays, which limit the current applied to the motors during acceleration. The engineer regulates the rheostat so that the current is kept to these limits as he pulls out of the spring switch at the east end of Eckman Yard. As the coal train works eastward into the upper part of the Elkhorn, the right-of-way twists and turns to follow the valley floor. Underway, precautions are taken should the power begin to build up on either the lead or the pusher locomotive. If the engineer sees that his engine is developing more power than is required under ordinary operating conditions, or that slack is running in or out, he immediately shuts off before the locomotive tractive effort builds up to an overload amount.

The coal train weaves through Keystone, past the 1800 ft. siding at Northfork and upward through Powhatan. At Elkhorn, WV, another 5180-ft. siding is available to accommodate traffic that sometimes stacks up on the hill ahead. Approaching a stop signal, the engineer on the lead electric locomotive reduces speed, which results in the load building up in the pusher. This is an indication to the pusher engineer that the engineer on the head end wishes to stop and he reduces his power accordingly. In this way the pusher keeps the train bunched, which helps when restarting the train.

The roadway becomes more restricted as the train passes through Ennis. Eastbound traffic is held to 14 mile per hour on account of the curvature. A 2500-ft. siding at Lick Branch is accompanied by a curve with another 14 mile per hour speed restriction. The coal train receives a clear signal and begins to pull upward over the last few miles of the Elkhorn grade. The-right of-way climbs above the valley floor and crosses over

LC-2 2512 rolls eastward near Bluestone Junction with a coal train. The electric locomotive fleet helped keep the railroad fluid during a period of exceptionally heavy traffic. *Collection of James N. Gillum*

the Little Fork of the Elkhorn, on the massive steel Bridge 681.

Break-in-twos on the hill are rare, but should they occur, the engineer on the lead locomotive attempts to release the brakes which have been automatically applied. The head-end engine continues to pull until stopped by the brakes on the train. The pusher engineer would promptly stop his section of the broken train. Another possible problem is a rare power failure. The load on the pusher engine is picked up car by car, until the current consumed by this locomotive reaches sufficient value to trip the circuit breakers. With planning, the engine circuit breakers are set to trip at a reasonable limit. This serves to greatly reduce the chance for a break-in-two under these conditions.

The electric coal train swings around a curve as it nears the final set of crossovers which are operated from Coaldale Tower. The tower, with the telegraph symbol CQ, controls the approaches to the single track Elkhorn Tunnel. Close co-ordination is necessary to get through the cramped bore. As the train works upgrade the crew keeps a sharp lookout for debris on the track. During the early 1940's intense mining activity in the immediate area posed an unexpected hazard to operation.

The mining is carried out along the coal out-cropping which parallels the right of way. At Elkhorn Tunnel, the coal seam is at track level. As the vein proceeds westward, the railroad right-of-way falls to below the coal before the track reaches Maybeury. Freezes and thaws during the winter of 1940-41 caused rock and debris disturbed during mining to settle and slide to the track. This posed a hazard to train operation as well as playing havoc with line side signal and power lines. Maintenance personnel were kept busy in both repair and patrol duties until the situation returned to normal.

Elkhorn tunnel is again approaching capacity. Eastward trains, at the tower operator's option, receive a signal to cross over from the eastbound to westbound track at the crossover located west of Coaldale in order to continue upgrade through the single track tunnel. This is used to maximize utilization of the tunnel by running trains around other traffic, which had encountered some sort of operating difficulty. The cramped conditions existing within Elkhorn Tunnel necessitates a Low Wire Warning. In the event of trouble inside the bore, timetable rules for electric coal trains instruct the electric current to be cut off in the event trainmen go on top of cars. A telephone is located in the middle of the tunnel that could be used for this purpose. The Conductor notifies the Power Director at Bluestone who has responsibility to cut the power. The current cannot be restored until the Power Director has been notified by the Conductor that men under his supervision were in the clear.

When the train reaches the summit at Ruth and begins to descend the grade, the load on the lead locomotive decreases as the train passes over the summit. When approximately half the train is over the summit the speed gradually increases until regeneration comes into action. The uncoupling lever between the caboose and the pusher locomotive is lifted. As the caboose passes over the summit it leaves the pusher behind. The pusher stops and drops back through the tunnel to the westbound track, and returns to the bottom of the hill.

Train speed is restricted to 14 miles per hour for eastbound movements at Ruth. Electric engines not regenerating are required to be handled in the same manner as steam-powered runs on the descent. Under these procedures, the train was first stopped and retainers are set. If the train was less than 125 cars and 5,000 tons, the retainers were turned up on 40 cars. If the train exceeded these limits, retainers were set on 50 cars. Whatever restrictions exist, they were still better than the "good old days" when the eastbound track used the old line down to the Bluestone River at Cooper. Speed limit for coal trains down the incline to Cooper was 12 m.p.h. where the infamous "Guard Rail Curve" was encountered. This was the worst spot on the line until track revision in 1923 eliminated the problem. As a safety precaution, the train line is checked in the caboose of all eastbound movements at Ruth and the value recorded.

Electric locomotives were sometimes called upon to push steam trains which had stalled or had other problems beyond Elkhorn Tunnel. At Coaldale Tower, if the pusher received instructions to bring the helper on through to Bluefield, operating rules required the crew to cut off at Ruth if the train line was not cut through to the pusher, in order to avoid shocks and slack action in handling the train. The pusher crew would follow the train east to HQ tower at Graham, then couple on to assist the train up the hill into Bluefield Yard.

Electrically-powered coal trains use regenerative braking over Bridge 854 at Cooper as they continue downgrade approaching Bluestone Junction. A permanently mounted slow board was erected at the west end of the train shed. Trains approaching this point were required to remain under control, prepared to stop in the event of track obstruction. Frequent movements to and from the Bluestone and Pocahontas branches were a source of delay, as well as motors exiting the inspection and maintenance area at the Bluestone Powerhouse. For westbound traffic, a 3-arm interlocking signal located on the east side of the Bluestone Powerhouse controls our coal train movement through the plant. Once past this point, the coal train contin-

The Pocahontas Division was a busy section of railroad, made worse by winter weather and wartime traffic. We look over the engineer's shoulder as we work our way eastward nearing Bluefield on the head end of a coal train.

ues three miles further eastward to Flat Top Yard, where all coal trains stopped to drop or fill out with additional loads.

Our train fills out with additional loads at Flat Top yard. A Class LC-2 electric locomotive is rated 5,800 tons from Iaeger to Tug and 4,200 tons between Tug and Eckman. From Eckman the rating to the summit is 2,100 tons. Coal trains usually depart Auville Yard with the maximum of 4,200 tons to avoid delays necessitated by tonnage reductions. Between Flat Top Yard and Bluefield, the rating per unit increased to 3,000 tons. A single set of LC-2 electric engines, which had brought 4,200 tons over the summit, could add another 1,800 tons before departing Flat Top yard for Bluefield.

The train is allowed a top speed of 28 m.p.h for the remaining seven and one-half miles into Bluefield. In heavy traffic conditions, special precautions were taken. One particular problem from the resulting difficulties of getting into Bluefield yard was the danger of running into traffic ahead. During foggy and inclement weather, crews in charge of steam locomotives eastward making less than 14 mph on the grade between HQ tower at Graham and Bluefield were required by rule to drop off a lighted flare at intervals for the benefit of electric trains. Flares burned for a regular interval of time, giving following traffic an additional warning of traffic ahead. Care also had to be taken in the presence of pushers that may be running ahead to assist trains that stalled on getting up the grade into Bluefield yard.

The inbound trains worked over the facing points of a spring switch, located at the pull-out switch for the Westbound Yard, upon entrance to Bluefield. If the train had been running wrong main against the current of traffic, the engineer had to watch for other traffic, observing the dwarf signal protecting this switch location before proceeding. Final Low Wire Warnings for the trip were located at the Allen Street and Mercer Street bridges in Bluefield Yard. The coal train slows as it pulls into a receiving track, guided by both signal and switchmen in the yard. The train is brought to a safe stop and brakes set on the loads. Our locomotives cut off at the area where the crew is relieved. ■

Glossary

A

A.C. - alternating current

B

b.t.u. - British Thermal Unit, a measure of heat

boiler - a container or apparatus used in the making of steam

C

catenary - the wire, structures and facilities used in the delivery of electricity

catenary bridges - structures which support the catenary at fixed points

coke - a hard gray porous fuel made by heating coal to drive off most of its volatile material

condensor pumps - a device used to apply force which changes steam vapor into liquid

condensation - the process of changing vapor to a liquid

cooling pond - a contained area in which water used in steam production disperses heat into the atmosphere thereby reducing the temperature of the water

current - A system by which A.C. power is delivered at one potential or voltage setting

cycle - a period of time occupied by a series of oscillations of polarity that repeat themselves regularly and in the same order

D

D.C. - direct current

dead-ending - a method of bracing utilizing a fixed position

draw - the bottom of a depression or contour in a land form

drift mine - a mining operation which enters a coal deposit horizontally, usually on the side of a hill or embankment

E

electric braking - a system of control in which the energy exerted by the force of a train on a grade is converted into electricity

F

feedwater heater - an apparatus that pre-heats water before introduction into a boiler, thereby reducing the amount of energy needed for its conversion to steam

friable - easily pulverized

H

herringbone reduction gear - a toothed wheel made up of rows of parallel lines with adjacent rows slanting in reverse direction that interlocks which another toothed wheel or shaft for transmitting motion

high tension switching equipment - apparatus used in the control of electrical current of high potential, or voltage

I

induction motor - a machine in which an electrical current is produced by the proximity of an electric or magnetic field

inductive interference - resistance in a transmission line brought about or amplified by the presence of a nearby electrical current

intake canal - an artificial channel through which

water is made available for use in the boiler

K

kilowatt - 1000 watts
k.v.a - kilovolt amperage

L

low-potential switching room - a partitioned part of a structure in which the apparatus used to control electric current of low potential or voltage is housed

O

1:2:5 concrete - a hard building material made of 1 part sand, 2 parts gravel and 5 parts cement

P

particulates - minute separate particles
powerhouse - a building in which electric power is generated

R

radial type brick - building blocks having sides arranged like rays coming from a common center point
regenerative braking - a system of control in which the energy exerted by the force of a train on a grade is taken up by the electrical motor on the locomotive and converted into electricity which is returned to the catenary
rotor - a rotating member of a mechanical or electrical device

S

seam - a layer of mineral matter
slack - loose or improperly sized coal
stator - the collection of stationary parts in the magnetic circuits of a machine or motor
step-up - a process by which the potential, or voltage of a current is increased
step-down - a process by which the potential or voltage of a current is decreased
stoker - a device which supplies fuel to a furnace
stoker engine - a mechanical apparatus used to provide power to a stoker

T

three-phase A.C - A system by which A.C. power is delivered in more than one potential or voltage setting, allowing a high torque to be used on grades and a high speed to be used under gradeless operating circumstances
transformer - a device which changes the potential or type of electric current
transmission - the passage of electricity between generating and receiving stations
transmission lines - the structures used in the process of transmitting
trolley wire - that portion of the electrical delivery system which provides contact with the locomotive

V

voltage drop - a fall in potential difference measured in volts, usually below a fixed standard

W

water tube boiler - a device of high thermal efficiency which converts water to steam in a process by which the water is introduced to heating through a series of tubular passageways

Index

V

W

X

Y

Z